RESTORING FARM WOODLANDS FOR WILDLIFE

David Lindenmayer, Damian Michael, Mason Crane, Daniel Florance and Emma Burns

RESTORING FARM WOODLANDS FOR WILDLIFE

David Lindenmayer, Damian Michael, Mason Crane,
Daniel Florance and Emma Burns

CSIRO

PUBLISHING

A catalogue record for this book is available from the National Library of Australia.

Published by:

CSIRO Publishing
36 Gardiner Road, Clayton VIC 3168
Private Bag 10, Clayton South VIC 3169
Australia

Telephone: [+613] 9545 8555
Email: csiropublishing@csiro.au
Website: www.publishing.csiro.au

Front cover: (main image) planting near Wagga Wagga (D. Blair); (thumbnails, left to right) crevice skink (D. Michael); sugar glider (M. Clancy); grey-crowned babblers (D. Smith). Back cover: (left to right) grey fantail (G. Chapman); plantings near Holbrook, NSW (M. Crane); swift parrot (G. Chapman).

Set in 11/13.5 Minion and Helvetica Neue
Edited by Joy Window (Living Language)
Cover design by Andrew Weatherill
Typeset by Thomson Digital
Index by Max McMaster
Printed by Ingram Lightning Source

Feb26_RP_ILS

Contents

Preface

'When I was a kid in the 1940s, if I had been naughty then dad told us to get outside and chop down a tree. Now it is the opposite – planting trees is important and there's far more vegetation than there used to be ...'

This quote is from Professor Ross Cunningham, an expert statistician who grew up near Walla Walla in the southern Riverina and whose statistical insights underpin many of the projects that feature in the chapters in this book. This quote highlights the positive changes in psyche, and the corresponding changes in the landscape, in agricultural south-eastern Australia. It also underscores how far we have come in a relatively short time. Much of that journey is a legacy of hard work on the ground by groups like Landcare and staff from regional organisations such as Local Land Services in New South Wales and Catchment Management Authorities in Victoria. At the same time, there has been an array of long-term research studies quantifying the environmental benefits of farm restoration. Much of this book outlines the results of work by our team of researchers and field-based scientists at The Australian National University.

Our sincere hope is that science and on-the-ground management will continue to intersect in positive and productive ways so that the integration of agricultural production and on-farm conservation are seen as the norm for any agri-business or other farming enterprise.

The authors
December 2017

Acknowledgements

This book is based on years of careful research that could only have been completed through enduring collaborations with scientific colleagues as well as with numerous farmers and skilled practitioners from natural resource management agencies like Local Land Services, Catchment Management Authorities, Landcare and Greening Australia.

There have been many generous funders of our work over the past two decades. These include the Australian Research Council, the Department of Environment and Energy, the Ian Potter Foundation, the Vincent Fairfax Family Foundation, the Calvert Jones Foundation, the New South Wales Environment Trust, the North East Catchment Management Authority, the Goulburn Broken Catchment Management Authority, Central Tablelands Local Land Services, Murray Local Land Services, Riverina Local Land Services, Southern Tablelands Local Land Services, and private donations from generous benefactors including Kent Keith and John Mitchell. Other major supporters have included Holbrook and Tarcutta Landcare. Karen Gair was instrumental in helping raise funding for our work from several important philanthropic institutions.

Many people based regionally have been strong supporters of our work for a prolonged period of time but particular mention needs to be made of Emmo Willinck, Tom White, Kylie Durant and Dale Stringer.

Many researchers have had significant input to our long-term farm work. These include Sam Banks, Philip Barton, Richard Beggs, Donna Belder, Clare Crane, Don Driscoll, Claire Foster, Phil Gibbons, Nicole Hansen, Karen Ikin, Geoffrey Kay, Adrian Manning, Rebecca Montague-Drake, Alessio Mortelliti, Nicki Munro, Kat Ng, Sachiko Okada, Thea O'Loughlin, Jennifer Pierson, Stephanie Pulsford, Chloe Sato, Ben Scheele, David Smith, Nici Sweaney, Ayesha Tulloch, Martin Westgate and Ding Li Yong.

The studies which underpin the content presented in this book could not have been instigated, analysed or completed without the assistance of outstanding expert statistical scientists, in particular Wade Blanchard, Ross Cunningham, Peter Lane, Jeff Wood and Alan Welsh.

None of the work presented in this book could have been completed without the help and encouragement of hundreds of individual farmers who allowed us access to work on their land and frequently provided deep insights into key, and often highly innovative, ways to tackle difficult restoration problems on their properties.

We thank Tabitha Boyer who assisted with many key aspects of manuscript preparation, and Clive Hilliker for the production of some graphics.

John Manger from CSIRO Publishing has championed the publication of our work as a series of books over the past 15 years and we are deeply grateful to him for his support and encouragement.

About the authors

David Lindenmayer is a Research Professor and ARC Laureate Fellow at The Australian National University who has specialised in established large-scale, long-term ecological monitoring and research programs in the temperate woodlands of south-eastern Australia. He has published more than 700 scientific articles and 45 books on conservation, restoration and natural resource management.

Dr Damian Michael is a Senior Research Officer in Ecology at The Australian National University. He has broad interests in landscape ecology, biodiversity conservation, herpetology and understanding the ecological importance of rocky outcrops in agricultural landscapes. He manages several large-scale biodiversity monitoring programs in New South Wales and has published 90 scientific papers and six books.

Mason Crane has been a field-based research officer with the Fenner School of Environment and Society at The Australian National University for the past 15 years. During this time he has implemented and worked across numerous research projects examining biodiversity conservation in agricultural landscapes. His main responsibility is to coordinate research programs associated with the South West Slopes Restoration Study. While having a broad interest in ecology and a wide range of taxa, Mason is in the final stages of a PhD program focusing on the conservation of the Squirrel Glider.

Daniel Florance has been a research officer with the Fenner School of Environment and Society at The Australian National University for the past 7 years and is responsible for field-based, long-term ecological research in south-east Australia. He has a strong interest in the conservation of our native woodland ecosystems, and using scientific research to provide practical evidence-based solutions to implement conservation within the agricultural landscape.

Emma Burns is a member of the Fenner School of Environment and Society. Currently she is the Director of Sustainable Farms, an ANU transdisciplinary initiative. She also sits on the Australian Ecosystem Science Council. Emma has a PhD in population genetics and phylogeography. She has held positions in consulting, government policy (NSW and Commonwealth), and research management. Emma has published on diverse topics in ecological research, conservation management and environmental policy.

'The best time to plant a tree was 20 years ago; the next best time to plant a tree is now.' (Anonymous)

1

Introduction

Millions of hectares of temperate woodland and billions of trees have been cleared from Australia's agricultural landscapes (Walker *et al.* 1993; Lindenmayer *et al.* 2010a). While this has facilitated the development of areas for cropping and pastures for livestock grazing, it also has had significant environmental impacts. These include rising water tables and secondary salinity, soil erosion, and losses of native plant and animal species. These environmental effects also have had significant negative impacts on farm-level productivity and ultimately profitability of farming enterprises. Restoring some of the native vegetation cover is one of the key ways to tackle these major environmental (and often financial) problems.

This book focuses on why restoration is important and describes some of the latest science about how to most effectively restore farms, primarily for different groups of animals, particularly birds, mammals and reptiles. Much of the focus of this book is on tree plantings, with reference to shrubs and undergrowth. This is for good reason; most problems with the loss of native plants and animals, rising water tables, increased soil erosion, and falling farm productivity correspond directly or indirectly with long-term losses in tree cover. Another focus of the book is on farm wildlife. This is because it has been the target of nearly 20 years of our research at The Australian National University on the effectiveness of restoration plantings (or replantings) for biodiversity. By far the most work we have done has been on temporal patterns of change in wildlife, and especially populations of birds. This has been the case because many farmers have a strong interest in birds, but also because birds can be coarse indicators of the recovery of other elements of biodiversity (Lindenmayer *et al.* 2014; Ikin *et al.* 2016a) (see Box 1.1). There is therefore a bias towards birds throughout this book.

Box 1.1. The call of the Clamorous Reed Warbler

'Warty-warty-warty-creep-creep-creep' is the call of a small and rather non-descript bird – the Clamorous Reed Warbler – that is closely associated with reeds and other vegetation at the margins of rivers, ponds, lakes and other waterways. The bird features in the title of a book on ecologically sustainable farm management by Charles Massey (Massey 2017). One of the reasons that the bird features in the title and in sections of the book is because its presence reflects, in part, the quality of waterways and water storage on a farm. This, in turn, is a coarse indicator of aspects of sustainable farming. That is, well-managed streams, dams and wetlands where farm management allows the recovery and maintenance of reeds and other riparian vegetation have a high probability of also supporting populations of the Clamorous Reed Warbler. Good management of 'the wet bits of a farm' is usually then a good reflection of the water-retaining ability of the land on a farm with corresponding benefits for livestock, particularly during drought periods. The Clamorous Reed Warbler is a seasonal migrant in southern Australia, arriving in September–October to begin breeding, so gauging the quality of habitat for the species requires surveys for the species at the appropriate time of the year. Of course, not all birds are good surrogates for farm management, and birds tend to be only crude indicators of the persistence of other groups of animals (such as reptiles) (Ikin *et al.* 2016a). Nevertheless, the distinctive call of the Clamorous Reed Warbler is one of the positive signs of good farm management.

In the last 5–10 years, there have been major new insights into what makes effective replanting programs. These insights have been derived at the same time as there has been a marked increase in the amount of native vegetation cover in some regions of south-eastern Australia, particularly central and northern Victoria and southern New South Wales. Some of these changes have been due to natural regeneration (e.g. post-drought recovery and post-fencing recovery along streamside areas), but other changes are a direct outcome of extensive and intensive revegetation programs. We have been fortunate to have been able to not only track changes in the amount of vegetation but also to document how biodiversity has responded to these changes.

There is no doubt that while revegetation programs have been important, new science shows there are ways to maximise their effectiveness. This book aims to bring some of that science together in one place. We have attempted to do this in a simple and sequential way. Our second chapter summarises some of the key reasons why restoration programs are important – why do a replanting? Chapter 3 examines what to plant. This is followed by a discussion in Chapter 4 of where to plant. Once a planting has been established, it will often need to be managed and this is the primary topic of Chapter 5. Nothing in nature is static, including plantings and in Chapter 6 our focus is on how plantings change through time.

Box 1.2. The potential consequences of doing it wrong – a key reason to write a book about plantings

One of the lessons from our long-term studies of plantings is that it is possible to establish plantings that can have a negative impact on biodiversity. As an example, the narrow plantings often one to three trees wide of exotic trees such as Cyprus Pine that have been established on some farms can actually create ideal habitat for exotic pest bird species such as the House Sparrow and European Starling. Native tree and shrub species are more likely to be colonised by native birds. We suggest it is far better to use native plant species, particularly those local to the broader region to establish plantings. Similarly, our field surveys show that very narrow plantings, especially those lacking an understorey and which are heavily grazed, tend to be dominated by the Noisy Miner (Fig. 1.1) – a despotic hyper-aggressive native species that bullies other birds (Grey *et al.* 1998; Montague-Drake *et al.* 2011; Mac Nally *et al.* 2012). Our work suggests that, rather than creating more habitat for this problematic species, wide plantings (with four or more trees) with some understorey will be a deterrent for the Noisy Miner and provide habitat for many other species of native birds (Lindenmayer *et al.* 2010b).

Figure 1.1. A narrow planting and the Noisy Miner. Planting photo by Damian Michael, Noisy Miner by Graeme Chapman.

1 Create a map of the farm including the natural assets

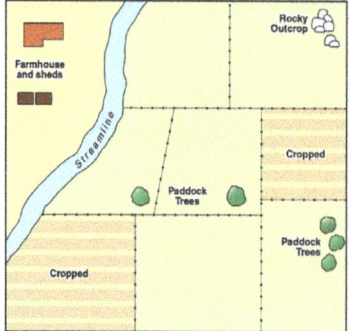

2 Write down the objectives of your management

- Protect paddock trees
- Encourage native birds
- Rabbit control

3 Consider the most appropriate and best options for management

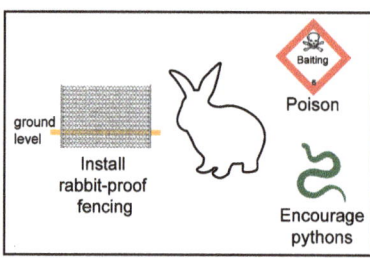

4 Discuss management options with neighbours and others with experience and expertise

5 Seek advice on what resources might be available to assist with planting, fencing, grazing control, and other actions

6 Record what has been done – including where, when and how

Date	Action	Status	Notes

Figure 1.2. Conceptual diagram of the sequence of steps in considering, designing, establishing and then managing a replanting.

The final chapter of this book explores some general themes, including the need for farm planning, and links between farm restoration and the importance of monitoring for documenting the effectiveness of restoration programs.

The chapters that follow are what we (and many colleagues) believe to be a logical sequence for a farmer or resource manager to think through the process of designing and then establishing a replanted area. The sequence of steps is shown in Fig. 1.2. This is a very linear way of approaching the challenge of establishing a replanted area on a farm. However, we recognise that a great many people have already been working to restore parts of the vegetation cover on their land and may benefit more from later chapters in the book, such as those on how to manage an already established planting or what to anticipate might be the changes in a planted area over time.

Where have the insights in this book come from?

Almost all of the information in this book is based on insights we have derived from the past 20 years of intensive field research. Our work aims to answer key scientific questions of applied management relevance, many of which are asked by farmers or staff from management agencies such as Local Land Services and Catchment Management Authorities. For example, how wide should a planting be to provide habitat for birds? Are older plantings more likely to be colonised by reptiles than young plantings? Do plantings with nest boxes support more species of native hollow-using mammals than plantings without nest boxes? A robust study design is needed to unequivocally answer these questions and considerable time is spent with statistical scientists ensuring that the work is completed at the highest possible standard. Once a study is designed, research and monitoring are conducted by expert field-based ecologists located permanently in regional Australia, and who complete many thousands of measurements on hundreds of hectares of plantings from central Victoria to south-east Queensland. The data that are gathered are then subjected to exhaustive and robust statistical analysis by a team of highly trained statistical scientists. The results of analyses of field data are then written up by all involved as scientific articles and submitted for publication to scientific journals where they are subjected to intense scrutiny under the peer review process (see Box 1.3).

There is some research in this book from our long-term studies which we have not yet published in scientific journals. An underplanting study within old growth woodland that lacks a shrub layer (see Box 3.2 in Chapter 3) is one of several examples. Our hope is that these studies will soon be finalised and the results more widely communicated to both our scientific colleagues and on-the-ground practitioners of restoration programs.

Box 1.3. Scientific 'peers', the peer review process and the quality of published science

The validity of scientific research is regularly 'road-tested' in the brutal world of scientific peer review. The concept of 'peers' in fields like politics is one in which a group of colleagues look out for each other, sometimes in somewhat less than ethically appropriate ways. Peer review in science is the extreme opposite when it comes to the assessment and publication of scientific articles. Reviewers work as hard as they possibly can to find flaws in work that is submitted to a good scientific journal, and they demand that the standard of research is lifted to as high a level as possible. Peer review is usually anonymous and criticism can be fierce; it is not a process for the faint hearted or those weak in spirit and persistence. Some journals reject at least 90% of articles that are submitted to them and a paper will often need to be started afresh when it is rejected. In the ecological sciences, the average paper might go to four or five journals before it is not rejected outright. Even then, a non-rejection may still require that the paper undergo major and repeated revisions before it is finally accepted. This process of submission, rejection, revision and acceptance may take two or three years and sometimes much longer. Almost all of the 100+ scientific papers that underpin this book have experienced this kind of brutal treatment in the peer review process. Peer review is a rigorous process that few non-scientists understand or appreciate, but, in general, it means there are relatively few poor quality articles in good quality scientific journals.

Box 1.4. Why science is important

A key part of applied ecological science is that it aims to better understand ecosystems in ways that lead to better land management. There are myriads of examples of why and how this can be important. A sobering but important example is that of the Wedge-tailed Eagle, rabbits and lambs. Hundreds of eagles used to be shot because they were perceived to be predators of lambs. Photographs of dead birds draped over barbed wire fences were common and one of us (DBL) still recalls this sight from early childhood. There is no doubt that lambs are sometimes taken by eagles. However, detailed scientific studies revealed that rabbits, and not lambs, are the main prey for the Wedge-tailed Eagle. Moreover, hunting of rabbits by birds of prey means less competition between sheep and rabbits for food, leading to better outcomes for farmers in terms of paddock productivity. In this case, robust scientific evidence highlighted the benefits of native wildlife for farmers.

Some of the other benefits of replanting on farms

While much of this book focuses on the responses of animals to replanted woodland vegetation, we are acutely aware of work on the value of restored areas in contributing to overall farm productivity and positive financial outcomes

(Walpole 1999; Bird *et al.* 2002; Massey 2017). There is also preliminary anecdotal evidence to suggest that there are positive relationships between restoration efforts and mental health; that is, farmers engaged in active restoration on their farms have found it to be beneficial to their mental health. We touch on some of these key (and sometimes somewhat unexpected) benefits of restoration programs in the final chapter of this book. Work on the intersection of restoration science, farm finance, and farmer well-being is complex, challenging and in its infancy, but it is our aim to report further on the results of that important multidisciplinary work in next 5–10 years (the amount of time it will take to confirm the multifaceted benefits derived from restoration programs on farms). This exciting new research is being enabled, in part, by the Sustainable Farms initiative at The Australian National University, launched in 2018.

Some notes about this book

Much has been written about revegetation and there are some excellent guides on the topic. We also have written extensively in the past on woodland restoration (e.g. Lindenmayer *et al.* 2011; Munro and Lindenmayer 2011; Lindenmayer *et al.* 2016c). Our aim here is not to rehash previous work so that we can lay claim to have written another book. Rather, a lot of new scientific work has been published in the past five years, resulting in the new ideas and results that we are featuring in this book. The majority of that work currently appears in scientific journals that few people read (including scientific colleagues) and which are simply inaccessible to people with a broad interest in restoration and wildlife on farms.

Communication of science is critical and through this book we have sought to create a way for practitioners with interests in establishing plantings in agricultural areas to access our research. In fact, one of the primary motivations for writing

---

Box 1.5. Things not covered in this book

The primary focus of this book is on farm restoration and particularly establishing (tree and shrub) plantings for wildlife, with an emphasis on work on birds, reptiles and mammals. We do not explore plantings established for commercial tree plantations for sawmilling and paper pulp production, farm forestry, and fruit and nut orchards. We also do not examine the role of plantings as shelterbelts in great detail (see Bird *et al.* 2002; Cleugh 2003), although in many cases well-designed plantings can have multiple important roles such as providing habitat for wildlife, creating shelter for livestock, and acting as long-term stores of carbon. Finally, we also do not explore how plantings might be best established to tackle problems with salinity (see Stirzaker *et al.* 2002). However, we recognise that plantings for salinity control also can make a significant contribution to biodiversity conservation if they are well designed.

this book was that many farmers and resource managers have repeatedly asked us over the past five or more years about our latest scientific insights about plantings. This book is a response to requests for more (and more accessible) knowledge about how to implement and manage effective vegetation restoration.

But there was another important reason to write this book – that is, a perception among some urban Australians that farmers are environmental pariahs destroying the nation's landscapes and biodiversity. There is no doubt that a small number of farmers do not do the right thing; unfortunate examples of extensive clearing and water theft are sad evidence of this. However, our experience over the past 20 years has been that many farmers are stewards of their land. Indeed, we have found very few farmers set out to deliberately destroy biodiversity. Rather, large numbers of farmers are doing excellent restoration work on their land. Biodiversity and indeed overall farm productivity have responded positively to these often very substantial management efforts. We therefore felt it was important to communicate to both rural and urban Australians that restoration programs can, and do, make a positive difference and that not all stories about the environment are invariably bad.

We are also acutely aware that urban Australians (who comprise more than 90% of the nation's population) are increasingly concerned about where their food and fibre come from and whether they are sustainably produced. Plantings and other kinds of restored native vegetation, and its associated biodiversity, are key parts of ecologically sustainable food and fibre production, and hence ecologically sustainable farm management. That is, farm management must be done in ways that do not further erode the not insubstantial losses of biodiversity that have characterised much of Australia, including agricultural landscapes. Sadly, a recent study suggests that only Indonesia has a worse recent record on biodiversity loss than Australia (Waldron *et al.* 2017). Ecologically sustainable farm management should therefore be seeking to improve biodiversity outcomes over time along with other environmental outcomes. Plantings are critical to achieving this outcome. Indeed, we view the so-called philosophical divide between farmers and those with concerns about environmental management as very much artificial and simplistic. Important Australian innovations such as Landcare arose from the combined efforts of peak farmer and conservation lobby groups. The same groups have come together to tackle issues associated with mining of coal seam gas.

Given that this is a science communication book rather than strictly a science book (although all of the content is underpinned by rigorous empirical science), we have kept referencing minimal, and the text short and to the point wherever possible. Longer and more in-depth discussion of key findings or major points can be found in the scientific articles that we have cited (using in-text references, also collected at the end of the book). If readers have difficulty in finding these articles, they can be obtained by writing to the authors.

Box 1.6. All that money raised from writing books

Make no mistake – very few scientists make money from books. Moreover, few scientists write books to make money (and they would starve if they tried). They do it to communicate their results, as we are attempting to do with this short volume. The limited revenue generated from books we have published so far has been reinvested in research, monitoring and outreach. Funds from this book will be no different.

A fundamentally important caveat

The how, what and where of plantings is driven very strongly by why. That is, the reason why a planting is being established will influence what is to be planted where, and how that planting is managed. The focus of this book is on establishing plantings for biodiversity conservation – that is, restoring largely cleared agricultural areas to enhance their value for biodiversity. We therefore reiterate that if the overarching objective is to do plantings for reasons other than promoting wildlife on farms, then some of the key recommendations that feature in our various chapters may not apply. Having said this, there are clearly many cases where, with some good planning, it can be possible to establish plantings that achieve multiple goals. For example, widening a shelterbelt so that a planting has both overstorey trees and an understorey can mean it plays a role in not only protecting livestock and improving lambing success, but also providing habitat for a greater range of native bird species than might otherwise have been the case (see Chapter 4).

Box 1.7. Stories of revegetation success

This book describes the many ways that plantings can be established and then well managed to create multiple benefits on farms. An increasing number of farms in south-eastern Australia have benefitted from establishing plantings, coupled with other aspects of sustainable farm management (see Massey 2017). One of these is Kincora farm near Gundagai on the South West Slopes of New South Wales. It was subject to extensive clearing in the 1920s and tens of thousands of trees were cut down. Over time, Kincora was subject to increasing problems with salinity, waterlogging of pastures, soil erosion, and a decline in paddock productivity. In the late 1980s, the owners of the farm implemented major changes in farm practices, including establishing hundreds of hectares of plantings, and revegetating waterways and farm dams. Problems with salinity and soil erosion have been fixed and the carrying capacity of formerly degraded but now restored pastures has increased. Long-term monitoring sites established by The Australian National University have recorded a major improvement in biodiversity on Kincora in the past decade. The farm is now more profitable than it has ever been and ecologically more sustainable. This is one of many success stories in which planting programs have played a significant part.

Figure 1.3. Archival images of Kincora farm near Gundagai where there have been significant positive benefits from major restoration programs including plantings. Provided by Sam Archer.

2

Why plant?

There are lots of reasons to establish plantings – the following chapters on how, where and when to establish plantings will often depend on these reasons and particularly on the objectives of the planting. The primary focus of this book is on the value of plantings to promote (and encourage) the conservation of wildlife on farms. However, there are many other reasons to establish plantings, and the second half of this chapter briefly touches on some of these other values.

Tackling biodiversity loss

Our research over the past two decades has repeatedly found that plantings are a critical part of efforts to conserve wildlife on farms, especially in heavily cleared and intensively managed landscapes. We have recorded more than 60 species of native birds and a more than a dozen species of reptiles in plantings during our repeated wildlife surveys on farms. Some of the key roles of plantings for biodiversity are summarised in Table 2.1. Perhaps most importantly, plantings form a different kind of habitat on a farm relative to other kinds of native vegetation such as natural regrowth and old growth woodland, and tend to support a different set of species (Lindenmayer *et al.* 2012a). This knowledge has been well documented in many of our long-term studies (Lindenmayer *et al.* 2016b) as well as by others working on wildlife on farms (e.g. Barrett *et al.* 1994; Martin *et al.* 2004; Barrett *et al.* 2008). Some of the bird species that prefer plantings include those of conservation concern like the Southern Whiteface, Rufous Whistler, Speckled Warbler and Flame Robin (Barrett *et al.* 2008; Cunningham *et al.* 2008; Lindenmayer *et al.* 2016b).

Box 2.1. Making Australia different from Europe

Farming landscapes in Europe have been subject to massive alteration over the past 1000 years. Only resilient species (and often those closely associated with humans) have survived in these heavily modified landscapes. Enormous losses of biodiversity have occurred in Europe. Many species disappeared from farming landscapes before people knew they even existed. Australia can avoid this same problem with conservation of existing remnant vegetation and major planting programs like those coordinated in the past few decades by Landcare groups, regional agencies such as Local Land Services, and Greening Australia. With good management, we can avoid the problems of massive biodiversity loss that have characterised agricultural landscapes in Europe.

Box 2.2. Our definition of habitat

The term 'habitat' is used frequently throughout this book. Plantings provide habitat for many species. Different species have different habitat requirements. A simple definition of habitat is:

The environment in which a species can occur, survive and reproduce.

Plantings can support several broad kinds of habitat including trees (both living and dead), understorey trees and shrubs, logs, rocks and groundcover attributes (e.g. native grasses), as well as topographical features such as rocky outcrops, creeks, rivers and wetlands.

Figure 2.1. The Flame Robin is an example of a bird species of conservation concern that is often found more often in plantings relative to other kinds of native vegetation on a farm (e.g. natural regrowth and old growth woodland). Photo by Donna Belder.

Table 2.1. A brief summary of some of the values of plantings for biodiversity we (and our colleagues) have demonstrated.

Biodiversity value of plantings*	References
Act as preferred habitat for some species of birds, including birds of conservation concern	Cunningham *et al.* 2008; Lindenmayer *et al.* 2010b; Lindenmayer *et al.* 2016b
Can exclude hyper-aggressive species like the Noisy Miner that depress populations of other native bird species	Lindenmayer *et al.* 2010b
Provide drought refugia for small and migratory birds	Lindenmayer *et al.*, unpublished data
Act as key areas for breeding by some species of birds	Belder *et al.*, unpublished data; Bond 2012
Provide foraging areas for the Squirrel Glider	Crane *et al.* 2014
Increase the cover around remnant old growth woodland and promote patch occupancy by arboreal marsupials	Lindenmayer *et al.* 2017b
Add to the range of habitats available to birds on farms, thereby promoting overall species richness	Cunningham *et al.* 2008; Lindenmayer *et al.* 2012a
Provide habitat for some species of reptiles, especially if grass tussocks are present and/or grazing regimes are not too intense	Michael *et al.* 2008; Pulsford *et al.* 2017a
Enhance connectivity for the movement of arboreal marsupials and assists them to colonise other areas of native vegetation	Lindenmayer *et al.* 2017b
Act as movement corridors for frogs	Pulsford *et al.* 2017b
Provide habitat for predatory insects that move out in adjacent crops and pastures (and potentially help control pest species)	Ng *et al.* 2017
Promote habitat suitability of farm dams for frogs	Hazell *et al.* 2001; Hazell *et al.* 2004
Enhance the protection of large old paddock trees	Lindenmayer *et al.* 2011; Lindenmayer and Laurance 2016
Help recruit new cohorts of paddock trees that are being lost at a rapid rate in farming landscapes	Fischer *et al.* 2009; Manning *et al.* 2013
Provide habitat for a range of species of invertebrates	Gibb and Cunningham 2010; Ng *et al.* 2017
Create 'reservoirs' of native plant species	Nurenberg, unpublished data

* Some of the values of plantings for biodiversity that have been identified by other researchers (such as a seed source for other plantings and for carbon storage) have not been listed in the table.

Biodiversity is nice to have around. Many scientific studies have shown that the physical and mental health of humans is enhanced in natural environments, including those that support populations of native plants and animals (Louv 2005; Zaradic and Pergams 2007). But, more importantly, biodiversity is necessary for ecosystems to function: the pollination of plants, cycling of nutrients and the decomposition of wastes (such as the carcasses of dead animals) (Barton *et al.* 2013). Research by The Australian National University has shown that replantings trigger the colonisation of farms and agricultural landscapes by a suite of different kinds of birds that would otherwise not be there. Other researchers have made similar findings (e.g. Martin *et al.* 2004; Barrett *et al.* 2008). These new birds bring different abilities to perform different functional roles in ecosystems. For example, plantings attract species that forage in the understorey, are migratory, and build cup and dome nests. Some of these birds, like honeyeaters, can eat up to 60% of the insects found on woodland trees. An Ibis can consume up to 250 g of pasture insects per day. These kinds of species can play important roles, such as consuming populations of insect prey. Other work has indicated that small-bodied birds in particular are attracted to the densely spaced trees in plantings (Fischer *et al.* 2008), possibly because they can more readily find food and nesting sites in these places. Plantings also may help them avoid predation by larger bird species. This analysis of bird body size relationships to plantings is important because many threatened woodland bird species are small-bodied animals (Montague-Drake *et al.* 2009). These results further demonstrate the very valuable conservation role that active restoration programs can play in promoting biodiversity conservation in agricultural landscapes.

Figure 2.2. Grey-crowned Babblers. This species is a charismatic species of conservation concern in south-eastern Australia. Increases in native vegetation cover in northern Victoria and south-eastern New South Wales through active restoration programs as well as natural regeneration have been important in contributing to the conservation of the species (see Robinson 2006). Importantly, our long-term data show this species has increased in abundance over the past 15 years. Photo by Dave Smith.

Box 2.3. Plantings as part of the 'portfolio of natural vegetation assets' on a farm

Much of this book is about plantings. Yet it is often hard to talk about plantings without also discussing other types of vegetation structure in the landscape, such as regrowth and old growth woodland. This is because while these different broad types of vegetation have habitat values for different elements of biodiversity, they also have combined value. That is, the biodiversity found on a farm responds not only to the range of habitats it supports, but also to the combined value of those habitats (Cunningham *et al.* 2008; Ikin *et al.* 2016b). This is often why the term 'portfolio of vegetation assets' is used in several parts of the book. As an example, this portfolio effect occurs during drought periods where plantings and regrowth woodland act as refuges from low rainfall and high temperatures for small-bodied and migratory birds. These kinds of birds are therefore better able to persist on farms during periods of environmental stress. These results suggest that the answer to the question 'Should I plant or should I let natural regeneration take place?' may sometimes not be an 'either/or' proposition. That is, where it is possible to have both, then this will promote higher overall levels of bird biodiversity on farms.

Box 2.4. There is no such thing as a 'perfect' planting

Throughout this book we discuss ways in which the value of a planting can be enhanced for biodiversity. However, we are acutely aware that there is no such thing as a 'perfect' planting for all elements of biodiversity. The densely spaced trees in some plantings will be good breeding habitat for some species of small woodland birds, but the changes in light and temperature regimes in these areas will not favour some native reptile species, especially those that seek out open, sunny habitats. The range of species, and differences in the habitat requirements of different species, presents an enormous challenge for those trying to integrate conservation and agricultural production. The solution to this problem is to ensure that the portfolio of vegetation assets is 'balanced'; that is, land managers create a range of conditions on a farm. In the case of plantings this can mean, for example, some old plantings, some young plantings, some dense plantings, and some more openly spaced plantings.

Plantings are also important areas for breeding by woodland birds (Barrett *et al.* 2008; Bond 2012) with new work at The Australian National University highlighting the wide range of species that breed successfully in these places (Belder *et al.*, unpublished data). In some cases, birds return repeatedly to the same planting to breed in successive years. This includes migratory species such as the Rufous Whistler and White-winged Triller that move large distances between seasons, including those that overwinter in far northern Australia and New Guinea. These insights come from individuals that are banded with unique combinations of coloured rings on their legs (Fig. 2.3).

Figure 2.3. Colour-banded birds caught in plantings in the South West Slopes of New South Wales. Unique combinations of colour bands allow researchers to determine which individual birds persist in plantings over the long term as well as those which return to plantings in successive years, often to then breed successfully. Photos by Donna Belder.

Other reasons to establish plantings on a farm

Because passive restoration methods like natural regeneration are not possible

Sometimes establishing plantings is the only way to revegetate parts of a farm. This is because the natural reserve of seeds stored in soil has been depleted and cannot be refilled naturally because of the absence of paddock trees or other kinds of plants that provide a source of seeds for new cohorts of trees or shrubs. In other cases, the loss of digging native small mammals, past clearing, prolonged high-intensity grazing, the predominance of weeds, severe secondary salinity, and the addition of fertiliser have altered soil and other conditions to the extent that natural regeneration cannot take place or may be very slow to develop (Dorrough

Box 2.5. Natural regeneration and reversing the decline of paddock trees

Large old trees are critical structures for farm wildlife (Manning *et al.* 2006). Many animals cannot survive without access to these trees – for nesting, foraging or both (Lindenmayer and Laurance 2016; Lindenmayer 2017). They are also really important stepping stones that facilitate the movement of animals throughout agricultural landscapes (Fischer and Lindenmayer 2002). Several studies have highlighted the substantial decline in paddock trees that is taking place in large parts of rural Australia (Manning *et al.* 2013). One way to address this problem is to alter grazing regimes to encourage tree regeneration (Sato *et al.* 2016). Reduced fertiliser application and short-rotation grazing with long rest periods (with each paddock grazing for less than 90 days per year) lead to the recruitment of new cohorts of trees that eventually may become large old paddock trees (Fischer *et al.* 2009). Paddock tree protection through fencing and targeted revegetation programs to encourage recruitment of new cohorts of paddock trees have gained increasing popularity with farmers and staff from management agencies such as Local Land Services and Landcare.

and Moxham 2005; Duncan and Dorrough 2009). In these cases, establishing plantings may be the best option that farmers and other resource managers have to increase vegetation cover. This can be a difficult decision because the direct costs of natural regeneration tend to be far cheaper than the direct cost of establishing plantings. Staff from the Australian Government estimated that the direct costs of replanting (in 2017 Australian dollars) in southern New South Wales were approximately $1700 per ha or ~5.5 times that of natural regeneration. These costs exclude fencing and the cost of the land that is taken out of production. However, areas where natural regeneration might be most feasible might be those that are least productive (Sato *et al.* 2016) and hence lost productivity may be relatively limited in these cases. In comparison, places requiring active planting may be on more productive areas (where the soil seed bank has been exhausted and trees will not establish through regeneration) and the indirect costs of revegetation through parts of paddocks being taken out of production may be higher.

To tackle salinity, rising water tables and water quality

Establishing plantings can be an essential step in efforts to tackle rising water tables and associated secondary salinity (Stirzaker *et al.* 2002). We have worked on many farms where plantings have been particularly effective in addressing these problems. Indeed, there has been a major increase in productivity on these properties with farmers now able to readily access previously waterlogged pastures. There also has been improved pasture grass growth, enabling increased rates of livestock production.

Figure 2.4. Plantings established in the Holbrook region of southern New South Wales specifically to tackle the problems of rising water tables and associated secondary salinity. Photo by Mason Crane.

Other landholders have used plantings to improve the quality of the drinking water for their livestock, for example through reducing soil erosion on their farms, particularly during floods. This has led to corresponding improvements in the health of the cattle and sheep, resulting in better prices for animals at market. At the same time, better management of streams, creeks, dams and wetlands has helped maintain more water for longer on farms and not only helped increase farm resilience during droughts, but also promoted farm wildlife conservation.

Box 2.6. The international reach of restoration work

This book focuses on Australia, in particular agricultural areas from northern Victoria and New South Wales to southern Queensland (a large, very productive and agriculturally crucial part of the nation). The international reach of Australian innovations, such as our approaches to restoration, has assisted many other countries. Zululand in South Africa has soil erosion, water quality and other related land degradation problems not unlike parts of the wheat–sheep belt of south-eastern and south-western Australia. People working on strategies for tackling some of these problems have sourced some of their inspiration from Landcare initiatives in Australia. Stabilising streambanks to limit erosion and maintain water quality has been one of the major land management tasks. Of course, like everywhere in the world, solutions to problems are often tailored in ways that make them as effective as possible. In the Zululand case, the answer was to move large rocks from the valley floors to steep and otherwise eroding stream channels – all by hand (see Fig. 2.5). Hundreds of channels and literally millions of rocks have been moved this way and water quality has been improved significantly as a result.

Figure 2.5. A restored stream channel in Zululand, South Africa. Photo by David Lindenmayer.

To promote productivity and farm profitability

Plantings can be established to provide shelter for livestock, including shade from extreme weather (Cleugh 2003). Shelterbelts reduce water consumption by livestock and enhance lambing success (Cleugh 2003). Livestock also gain weight more quickly when they are in pastures protected by shelterbelts (see Box 2.7). Plantings also can support populations of beneficial insects (Gibb and Cunningham 2010; Thomson and Hoffmann 2010a; Thomson and Hoffmann 2010b) such as those which are pollinators of crops like canola and insects which are predators of crop and pasture pests. Trees and shrubs used in windbreaks can reduce evapotranspiration and limit soil erosion; crop yields are often higher on the leeward side of plantings (Bird *et al.* 2002). The vegetation in plantings can add to farm production in other ways. For example, plantings can be a source of food for livestock during times of low pasture growth. However, our long-term work indicates that the intensity of grazing needs to be managed in plantings, otherwise the benefits of such restored areas can be undermined (Lindenmayer *et al.* 2018) (see Chapter 5).

Increased financial value of the land

The preceding section very briefly highlighted the range of roles that plantings can play in enhancing farm productivity. These benefits can result in farms which support plantings being more profitable (Walpole 1999), including during drought periods when many farm enterprises can be under considerable financial stress

Box 2.7. Imagine

Imagine you are standing in a paddock. There are no trees and you are naked, not even a hat on your head. It is raining and the wind is blowing a gale. You are soaked to your (naked) skin. After 10 hours of driving rain and wind, the weather clears. Now the sun is out and it's baking hot. The rays are beating down. You are still in the middle of the paddock, with no trees for shade and protection. Now imagine what it is like to be a cow or a sheep. Farmers who truly understand animal production as well as animal welfare fully recognise that the protection provided by trees and shelterbelts more than outweighs the limited amount of pasture foregone to establish them.

(see the many testimonies in the book by Massey 2017). There is also some evidence that, in some regions, farms with reasonable levels of native vegetation cover (including plantings) have a higher land value and fetch a higher price when sold (Walpole *et al*. 1998; Lindenmayer 2017).

For carbon storage

Carbon is one of the primary components of trees. Trees are an excellent way to store large amounts of carbon in the long term. Indeed, long-term carbon storage will be essential in efforts to tackle the major problems of rapid climate change facing human societies (Mackey *et al*. 2013). The large and long-lived trees in Australian temperate woodland environments are often characterised by very high levels of wood density. In addition, Australian trees often have very high levels of biomass relative to the amount of rainfall that occurs in a given ecosystem. These

Box 2.8. Being paid to plant

Many schemes worldwide pay farmers to undertake conservation work on their farms. Agri-environment schemes in North America and Europe make billions of dollars in payments to farmers every year (Whittingham 2007; Michael *et al*. 2014; Batáry *et al*. 2015). Australia also has various forms of agri-environment schemes. An example is the Australian Government's Box Gum Grassy Woodland Project within the Environmental Stewardship Program, which extends from southern New South Wales to south-east Queensland and encompasses more than 150 productive farming enterprises (Lindenmayer *et al*. 2012c; Burns *et al*. 2016). Establishing plantings is one type of management for which farmers in the Australian Government's Box Gum Grassy Woodland Environmental Stewardship Program are paid under long-term contracts (up to 15 years) to conduct ecologically sustainable farm management and improve the condition of Box Gum Grassy Woodlands and biodiversity in agricultural landscapes.

Box 2.9. Plantings, finances and farmer well-being – the Sustainable Farms initiative

Farming is often a tough and often lonely business. It is no secret that there are major issues with the mental health and well-being of people on the land. Over almost 20 years of working with farmers, it has become clear that the mental health of farmers who have undertaken restoration works on their farms appears to be better than those who have not. The reasons for this remain unclear. Peter Herriot is a classic example. His farm is located near Holbrook in southern New South Wales and has suffered major problems with water logging, with many paddocks impassable in some years. Even access to his shearing shed was difficult in the years when surrounding areas were constantly saturated. After a visit to a set of now famous Potter farms in western Victoria, Peter Herriot was inspired to embark on a program of revegetating key parts of his property. The improvements he achieved in farm productivity and profitability were profound. But perhaps more important is the way Peter Herriot talks about how the planted areas on his farm provided a quiet place for him and his family members to think and rejuvenate, particularly during tough times. The plantings also provided a tangible sense of real achievement that resulted in a major improvement in the land, a testament to good farm stewardship (see https://vimeo.com/177640227). Our new work on farms and farmers aims to determine the reasons for the links between good farm management and better farmer mental health and well-being.

factors mean that plantings, and other wooded vegetation, have the potential to store very large amounts of carbon over prolonged periods of time. However, for farmers to gain a financial benefit from carbon storage, a coherent and transparent carbon market needs to be established, and unfortunately this has yet to eventuate in Australia. If such a market were to be developed (and we strongly suggest that it needs to be – and soon), the carbon storage could be an additional form of income for farmers.

Farmer mental health

One of the most unexpected consequences of our long-term research has been anecdotal indications that many farmers who undertake restoration work are in a better mental health state than those who do not (see Box 2.9). The reasons for these (admittedly preliminary) findings remain unclear. It might be because plantings provide farmers with a source of satisfaction through having achieved something very positive for their land. Or it may be that being part of a network (such as Landcare) provides valuable social interaction with other members of the community. Establishing plantings may promote the mental health of farmers because they can add scenic value to a property or increase the number of native animals that occur on a farm; many farmers tell us that they greatly enjoy the array

Figure 2.6. Peter Herriot on his farm near Holbrook in southern New South Wales. Photo by James Walsh.

of bird species that occur on a well-managed property that they have restored. Improved mental health could also be due to the results of such work, where the land will be in better shape than it was when they first started farming. It may be that plantings help productivity and profitability, and this eases financial stress (with its massive attendant impacts on mental health). Any or all of these reasons will be relevant and we will endeavour over the coming decade to better determine why it is that farmers with restored areas on their farms appears to be more mentally resilient.

Summary

There are many reasons to establish plantings on a farm and this chapter touches on some of these. They include tackling problems with farm productivity and profitability, storing carbon and generating alternative sources of income (such as through stewardship payments). Tree plantings allow landholders to establish vegetation where they want to on a farm and to meet particular objectives. In some cases, planting is essential since natural regeneration is not possible because of the depletion of the soil seed bank or changes in soil and other properties that prevent regeneration from occurring. While the cost of establishing plantings may seem like a large initial outlay, there is strong evidence that successful revegetation will not only increase productivity and profits, and lead to positive biodiversity and other environmental outcomes, but may also lead to improved farmer mental health.

3

What to plant? The content of plantings

Having decided to embark on a restoration program, the next challenge is determining what to plant. As emphasised earlier in this book, what is planted will be strongly influenced by the objectives of planting – different kinds of plants might be established if the primary aim is setting up shelterbelts as opposed to the aim of plantings being to provide habitat for native wildlife. This chapter focuses on the kinds of plants to establish when the objective is to restore habitat for wildlife, particularly birds, mammals and reptiles. This chapter is short as decisions about what to plant should be relatively straightforward.

Use native plants

All plantings should be established with Australian native plants. There are some good reasons for this. First, exotic tree and shrub species can become weeds. Willows in streambanks are one example and weeds like this can, in turn, compromise habitat suitability for iconic native animals such as the Platypus. Second, non-native plants can attract non-native bird species such as the House Sparrow, Goldfinch, Blackbird and Common Starling. Sometimes particular native tree and shrub species may need to be established to meet the requirements of particular target animals. For instance, She-oaks (Casuarina species) within plantings can be important for providing food for the Glossy Black-cockatoo. Third, native plants attract more species of native invertebrates that include

Box 3.1. Select the 'right' plants for the right places

Different species of plants grow best under different conditions. Some grow better on ridges whereas the best environments for others are in gullies. Matching plant species to environmental conditions (e.g. soil and rainfall conditions) has a significant effect on rates of growth and mortality, and therefore ultimately the success of restoration programs. Such matching requires specialist knowledge and information from experienced staff within Landcare, Greening Australia, and regional staff from natural resource management agencies. Increasingly, we need to choose plant species that have high tolerance to environmental extremes. Again, expert guidance is needed and is freely available in Australia.

predators of pest insects, crop pollinators and prey for native animals (Majer *et al.* 2001; Gibb and Cunningham 2010; Thomson and Hoffmann 2013). Four, incentive schemes that pay farmers to do revegetation on farms will always (and quite rightly) demand that native plants are used. Fifth, native plants are more likely to regenerate naturally (including after fire) than exotic plant species.

In general, it is best to ensure that a range of plant species is used to establish plantings. This is because the different kinds of flowers, seeds, bark micro-environments and other characteristics of different plant species ensure there is a range of resources available for a greater range of animal species, thereby boosting the biodiversity benefits of a planting. It also ensures resilience of plantings to environmental conditions, for example insurance against pest attack, disease and drought, and increasing frost resistance, by spreading the risk. Planting a range of species also adds different structural and age components to a planting. For example, it can include relatively short-lived wattle species that eventually create important dead timber among other longer lived eucalypt species.

Most guide books for establishing plantings recommend the use of local provenances of native plants. We understand this recommendation because many species are likely to survive best when they are well adapted to local conditions. However, sometimes there may be value in establishing native, but non-local, plants. For example, we have witnessed strong positive responses of declining native bird species such as the Crested Shrike-tit to non-local trees such as Ribbon (Manna) Gum and Tasmanian Blue Gum in southern New South Wales. This is likely because the bark streamers associated with these tree species provide high-quality foraging habitat for the Crested Shrike-tit. Other species such as the Marbled Gecko also benefit positively from the establishment of tree species with bark streamers because they produce suitable bark habitat faster than some local tree species. Winter-flowering species such as Yellow Gum, Mugga Ironbark and White Box provide important food resources for spectacular native animals like

Figure 3.1. Plantings that have some winter flowering species of trees can be valuable food resources for spectacular but now rapidly declining threatened animals like the Swift Parrot. Photo by Graeme Chapman.

the Squirrel and Sugar Gliders as well as many migratory and nomadic bird species. Non-local species such as Lemon-scented and Spotted Gums provide a prolific number of flowers used by the Grey-headed Flying Fox, Squirrel Glider, and a myriad of pollinating insects.

Figure 3.2. Ribbon gum tree and bark streamers provide valuable foraging habitat for species such as the Crested Shrike-tit. This spectacular bird species can sometimes be heard well before it is seen, not only because of its call, but also because of the noise it makes when stripping bark while hunting for insects and other invertebrates. Photo by Mason Crane.

Create an understorey or shrub layer where possible

Most of the focus of past planting programs has been on establishing overstorey trees such as eucalypts. More recently, many restoration practitioners have recognised that an understorey is an important addition to planted areas. For example, an understorey of Acacia trees and shrubs can limit the suitability of plantings for the hyper-aggressive Noisy Miner (Lindenmayer *et al.* 2010b), a bird species which can drive away many native bird species, particularly small-bodied birds (Montague-Drake *et al.* 2011; Mac Nally *et al.* 2012). Plantings with an understorey have more layers of vegetation and this provides more niches in which birds can forage and nest (Gould and Mackey 2015). This explains why plantings with an understorey typically support higher bird species richness than those without an understorey. Plantings with more plant species, including those that comprise the understorey and other layers (such as the ground layer), are also characterised by more species of insects, including predatory and herbivorous insect species (Ng *et al.* 2018). These insects not only provide food for other animals such as birds, frogs and reptiles, but also may have key ecological roles such as pollination and pest control, including in adjacent pastures and croplands.

The understorey of plantings, particularly Acacia trees and shrubs, also can be a valuable food resource for species such as the Squirrel Glider and the Sugar Glider, which cannot rest and den in young replanted areas but can use them for feeding. Many Acacia species produce suckers from their root systems creating new habitats, whereas others senesce at an early age, thus providing an additional habitat in the form of dead material and the accumulation of litter.

Box 3.2. An experiment to establish an understorey

Many areas of temperate woodland lack on understorey. This is also the case in many plantings. Lack of understorey may explain the absence of many species of small woodland birds that nest and feed in these parts of woodland environments. Re-establishing an understorey can be difficult where years of intense grazing have depleted the soil seed bank and only overstorey trees grow. In the early to mid-2000s, we instigated an underplanting study in which Acacia trees and small shrubs were deliberately established below Box Gum woodland trees. At the same time we commenced reptile and bird monitoring and the work showed that it typically requires four to seven years for small bush birds to begin to colonise these places. Populations of hyper-aggressive Noisy Miners continue to persist in these unplanted woodlands, but small woodland birds are able to coexist with them – possibly because they are not as readily seen by miners (Lindenmayer *et al.* in press).

Figure 3.3. The Sugar Glider (left) and Squirrel Glider (right). These two species are strongly associated with Acacia trees, in part because they consume the gum but also because of the insects associated with these trees. Sugar Glider photo by Matt Clancy, Squirrel Glider by Katherine Tuft.

Box 3.3. Keeping out partners in crime – why the understorey of plantings is critical

The notorious role of the Noisy Miner as a despot infamous for aggressively chasing other birds away is well documented. The role of plantings and particularly plantings with an understorey in suppressing populations of the Noisy Miner is increasingly appreciated (Lindenmayer *et al.* 2010b; Lindenmayer *et al.* 2016b). What is only beginning to emerge is that the Noisy Miner is often associated with the Grey Butcherbird – a highly effective predator of other birds and nests. Our long-term data show that these two species almost always co-occur in woodlands. Fortunately, both species are relatively rare in plantings, possibly one of the reasons why some species of small birds are more likely to occur within (and also breed more successfully in) plantings.

Figure 3.4. Relationships between the occurrence of the Noisy Miner and the amount of wattle (Acacia) in the understorey of a planting. Interestingly, populations of the Noisy Miner have undergone a significant decline in the South West Slopes region of southern New South Wales over the past 15 years, possibly as a consequence of increasing amounts of planting and natural regeneration on many farms. Photo by Graeme Chapman.

Figure 3.5. The Noisy Miner and the Grey Butcherbird. Noisy Miner photo by Graeme Chapman, Grey Butcherbird by Dave Smith.

The ground layer

The ground layer of plantings can be important for many native animal species. Some species of ground-foraging birds can be absent from plantings that lack forbs in the ground layer (Barrett *et al.* 2008). Parts of the ground layer such as native tussock grasses provide habitat for reptiles. Logs are also an important part of the ground layer of plantings (Selwood *et al.* 2009) and bird species richness is higher where these key elements of vegetation structure are present (Lindenmayer *et al.* 2010b). The paucity of logs within plantings may be one of the reasons that species such as the Brown Treecreeper are generally absent from plantings (Martin *et al.* 2004). Logs also can be critical habitat for reptiles and invertebrates such as a myriad of beetle species (Barton *et al.* 2011) as well as earthworms. Large logs are absent from young plantings as it takes considerable time for trees to develop large lateral branches or trunks that can eventually become large pieces of woody debris on the woodland floor (Manning *et al.* 2007). Some farmers have taken steps to rectify this problem by using a tractor and cable to tow fallen timber from cropping paddocks into plantings. Others have worked hard to avoid the temptation to tidy up fallen timber on parts of their farm (see Box 5.11 in Chapter 5).

Planting density

What is the right spacing of plants for a planting? It depends on what objectives and outcomes are desired. For example, for a shelterbelt, trees may be closely spaced to slow windspeeds (Cleugh 2003). Similarly, closely spaced trees are beneficial where the aim is to tackle problems with secondary salinity (Stirzaker *et al.* 2002). If the aim is to reduce soil erosion, trees may be widely spaced so that a ground cover of native plants can be maintained. Even when the primary

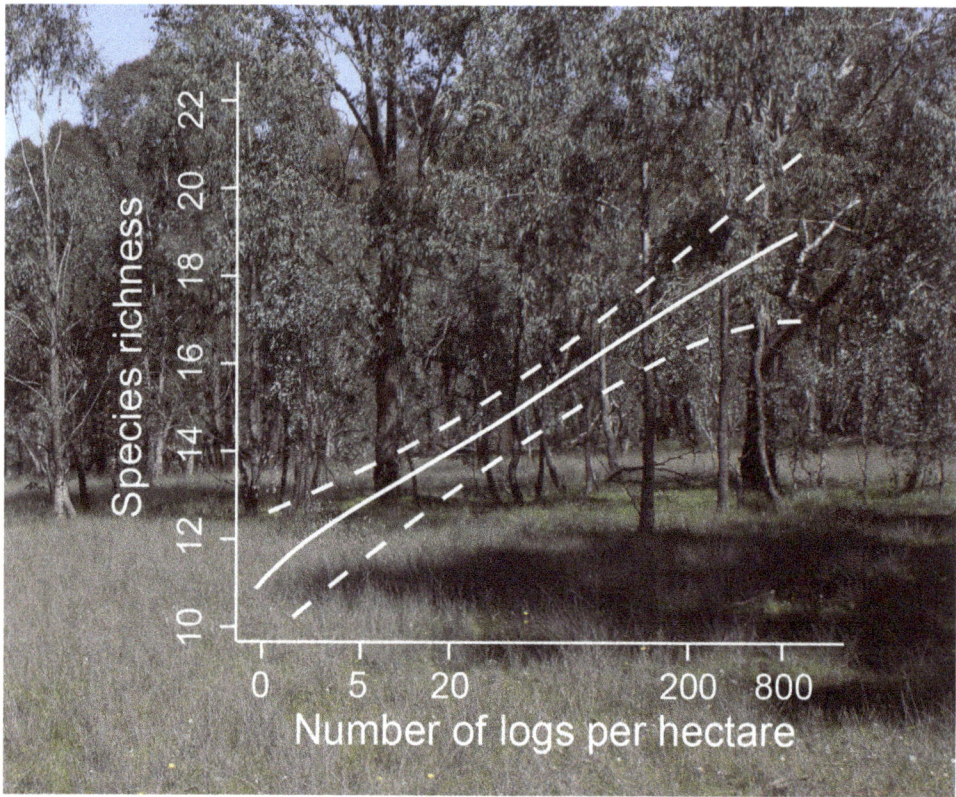

Figure 3.6. The relationship between the number of bird species and the number of logs within a planting (based on data in Lindenmayer *et al.* 2010b).

Box 3.4. Ground layer establishment of rare plants species

Many native ground cover plants are becoming increasingly rare in remnant woodlands. Through our work with the Threatened Species Recovery Hub (part of the Australian Government's National Environmental Science Program), together with our colleagues from Greening Australia, we have begun trialling a planting intervention to establish a suite of grazing-sensitive native forbs including Bulbine Lily, Yam Daisy and Variable Plantain. The plan is to expand this to include endangered wildflower species such as the Yass Daisy and Button Wrinklewort within existing woodland patches. A key part of this experiment is to test the establishment rate of rare and declining native plants under different grazing regimes to inform future planting and restoration works.

Figure 3.7. The Yass Daisy is a grazing sensitive forb that the Threatened Species Recovery Hub is hoping to re-establish. Photo by Mason Crane.

objective is to establish a planting for biodiversity, there is no 'one size fits all' recipe. This is because different species are associated with habitats of different density and spacing. However, a general rule is to try to ensure some variability in the spacing of plants across a planting – some areas that are densely spaced and others where the trees and shrubs are more widely spaced. Such variation in planting density might be within the same planting or between different plantings.

The density of plantings also will vary depending on whether trees or shrubs or a combination of both are being established. A planting densely planted with small to medium shrubs will be very different from one densely planted with trees. In general, smaller shrubs and ground covers should be planted at high density to compensate for the high level of mortality that is likely to occur. This also helps to promote a well-stocked seed bank to ensure that species persist in the longer term (including as seeds in the soil).

Important additional key structures in plantings

Plantings can often lack some key structures that are an important part of the habitat of species. Large old trees are often absent from plantings (Vesk *et al.* 2008) unless such areas are deliberately established around existing paddock trees and areas with scattered trees. When large hollow-bearing trees are absent from plantings, nest boxes can be installed to encourage colonisation and use by some cavity-dependent species. Nest boxes need to be well designed to meet the requirements of target species (Goldingay *et al.* 2015) and avoid problems such as boosting pest species like the European Starling (Lindenmayer *et al.* 2016a). We further discuss issues with the management of nest boxes within plantings in Chapter 5.

Other important structures within plantings that can have a significant positive effect on their value as habitat for wildlife include clumps of mistletoe, the presence of dead trees and shrubs, bush rocks and small rocky outcrops. Mistletoe is used by many bird species and some possums and gliders as a food source (Reid 1991; Watson 2001). Plantings with mistletoe have higher bird species richness than those where clumps of these parasitic plants are absent (Lindenmayer *et al.* 2010b). Dead trees provide nesting sites for some reptiles (such as goannas), mammals (such as bats) and birds (like the Dusky Woodswallow, which makes nests under strips of bark on dead stems). Rocky outcrops support very high levels of species richness of reptiles. The number of species of reptiles, as well as the abundance of species like the Southern Rainbow Skink, is elevated in plantings where rocky outcrops are present (Pulsford *et al.* 2017a), and also where the spacing of trees is not so high that incoming levels of light and solar radiation are impaired (Michael *et al.* 2010).

Figure 3.8. Southern Rainbow Skink, a species that can be found in relatively high numbers in plantings, particularly when they are not subject to high-intensity grazing by domestic livestock. Photo by Mason Crane.

How to plant – tubestock versus direct seeding

Many excellent books and scientific articles have been written on how to establish restoration plantings and we do not intend to repeat the extensive and excellent detail in those previous volumes (e.g. Greening Australia 2003; Streatfield *et al.* 2010). There are two basic ways to establish a planting after the site has been prepared (e.g. by spraying weeds and grasses and deep ripping the soil): from tubestock or by direct seeding. The most appropriate method will depend on factors like rainfall, past site history, topography and soil type. The best advice on which one to choose will be based on local knowledge from other farmers and staff from natural resource management agencies as well as groups such as Landcare and Greening Australia.

Tubestock are seedlings grown in a nursery from seed or from cuttings. Individual plants established from tubestock usually (although certainly not always) have a high rate of success compared to direct seeding. In general, tubestock is most appropriate for steep and rocky areas where access by seeding machinery is difficult. However, careful site preparation before the use of tubestock can be needed, especially if there has been extensive past soil erosion and if ripping is planned.

Tubestock are usually more costly than direct seeding, in part because each plant has to be individually hand-planted. Tubestock will therefore often be prohibitively expensive for large plantings. However, direct seeding may not always be very effective or logistically possible, especially in areas with a high cover of exotic perennial grasses or where there has been prolonged use of fertiliser and high-intensity grazing, leading to high levels of nutrients that are not conducive to the growth of eucalypts. In some cases tree guards will be required to protect young plants from being browsed by kangaroos and feral herbivores such as rabbits and hares. Tree guards also help protect young trees and shrubs from frost and desiccation. The local threats to planting establishment will inform the types of guards required. For example, cartons or plastic sleeves may be suitable for frost protection and herbicide spray drift, but may be insufficient where herbivore grazing is the main factor affecting plant survival.

Box 3.5. Before you plant

It always seems like it is good to get things started on the ground – to be doing something practical. However, it is important to plan ahead. First, check to make sure there are no buried or overhead utilities (such as telecommunication, gas mains and power lines; see https://www.1100.com.au), or easements where you cannot plant. Keep in mind that electricity companies may not appreciate having to trim planted trees from under power lines in 20 years' time. If planting in the vicinity of utilities, think about planting shrubs that won't need trimming in the future and plan an access point in case utility companies need access. Second, it can be useful to assess long-range climate forecasts to assist timing of planting as this helps reduce rates of tree mortality. Third, if a planting is to be established using tubestock, then ensure that shrubs and trees are ordered early to guarantee supply and allow planting to occur at an appropriate time of the year (usually late winter or early spring). Fourth, consider the location where the planting will be done and make sure that native grassland is not being destroyed to establish a planting. Establish fences before a planting is commenced so that young plants are protected from grazing by livestock. If direct seeding, it is important to ensure there is adequate access for machinery. The area targeted for planting also may need to be subject to weed control well in advance of establishing native plants. Various methods of weed control are available and, again, need to be matched to local conditions. They include boom-spraying, spot-spraying, scalping, the use of weed mats, and sometimes intensive grazing. Prepare the soil before planting, for example by deep-ripping the soil. However, if an enhancement planting is being established under or close to existing trees, don't rip under the drip line of existing trees. Also do not rip across drainage lines or in areas that have established biodiversity values (e.g. swards of native grass, patches of native wildflowers).

Figure 3.9. Tubestock planting process. Photo by Mason Crane.

Direct seeding is the other popular method in which plantings are established. In this method, seeds are spread directly on the ground, either by hand or purpose-built machines (or occasionally by air). Sometimes direct seeding can be done by laying down branches with the seed attached. Rates of germinant success tend to be lower with direct seeding, but the method is often more cost-effective than tubestock (although large amounts of seed needs to be gathered so may not be suitable for rarer species or those species for which seed is not readily available). In addition, the spacing of trees and shrubs is less regimented compared to the straight lines of plants that characterise most tubestock plantings. Sporadic, staggered germination of seed can create a more 'natural' look to a planting. The longevity of some native seed species (several years for many Acacia and eucalypt species) means direct seeding can be undertaken in drier conditions and germinate as conditions allow. Indeed, many successful seeded plantings have been sown during droughts. Quite sophisticated approaches (and associated machinery) have been developed for direct seeding. For example, some single-pass mechanical seeders can herbicide spray, scalp, rip the soil, sow the seed (at different depths and with different inoculants such as smoke water), and then cover the seed with soil. Weed control before direct seeding is an important consideration because the precise location of young plants is not known, making it challenging to target control efforts like herbicide application. Single-pass seeders allow for spraying of herbicide as the seed is sown, so can require less site preparation. However, experience suggests that in sites dominated by an exotic understorey, seeding success may be less likely without some weed management before seeding, and may require several successive sprays in advance to reduce weed load and competition. Consideration needs to be given to the location of the proposed planting as direct seeding may not be appropriate in very rocky and/or steep areas (these areas may be better suited to tubestock or hand-seeding).

Excluding grazing by domestic livestock is important in both tubestock and direct seeding plantings, otherwise young plants are eaten or trampled and killed and important characteristics of plantings such as the leaf litter layer are damaged or lost. Fencing is the logical way to manage grazing pressure in quarantine planting. Our field data suggest that the longer high-intensity grazing is excluded from plantings, the higher the number of species of birds and reptiles these areas support (Lindenmayer et al. 2018). As discussed elsewhere in this book, grazing and browsing by other herbivores, both feral and native (e.g. rabbits and kangaroos) can damage plantings and severely impair their development, and populations of these animals may need to be controlled.

Figure 3.10. A 'parent' Blakely's Red Gum ringed by seeded new regrowth 'children'. Mulligan's Flat Nature Reserve, Canberra, Australian Capital Territory. Photo by Esther Beaton.

Summary

In general, plantings that are better for biodiversity will be those that:

- are established with native plants
- have a native understorey and ground cover
- are established around other key structures such as large old trees, fallen timber and rocky areas.

Tubestock and direct seeding are the two major forms of establishing plantings. The success of both methods is dependent on pre-establishment site preparation (such as weed control and creating a viable soil seed bed), appropriate fencing and the control of grazing pressure (from livestock, feral animals and native herbivores). Assessing long-range weather forecasts also can help reduce rates of mortality due to weather extremes, especially when plants are young.

4

How much and where to plant? The size, shape, location and surrounding configuration of plantings

Chapter 2 explored some of the many reasons why replanting native vegetation is important. The preceding chapter (Chapter 3) provided information on what to plant. The focus of this chapter is on how much to plant, the shape of plantings, and where to establish plantings – that is, which part of the farm is appropriate and why. The material we present encompasses topics such as the size and shape of plantings, and the influence of the adjacency of other native vegetation on the effectiveness of plantings for some elements of the biota.

How much to plant

For several years, scientists considered that native vegetation needed to exceed more than 30% of a farm or a landscape to conserve biodiversity. This was based on theory and the basis of the estimates seem plausible, although empirical evidence to support this 30% 'threshold' value was lacking. Values of 30% native vegetation cover would clearly be very hard to achieve for many farmers, landscape managers and resource management agencies. Our work indicates that landscapes and farms with higher levels of native vegetation cover do indeed support more species of groups like birds (Cunningham *et al.* 2014a; Cunningham *et al.* 2014b). However, many farms with lower levels of cover support large numbers of species

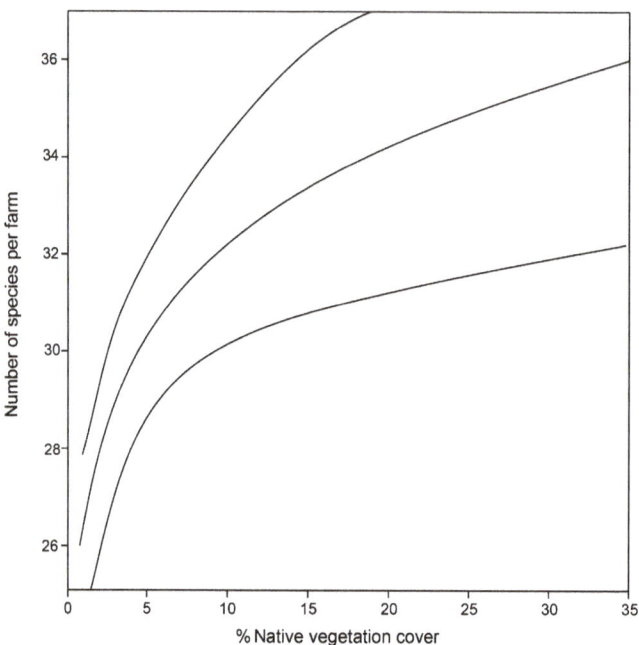

Figure 4.1. The diminishing returns relationship for overall bird species richness (Cunningham *et al.* 2014b).

of not only birds but also species of other groups such as reptiles and plants. Part of this research suggests that farms even with relatively low levels of cover can be important for biodiversity. This work also shows that strong gains in bird biodiversity can be made with restoration programs that add more native vegetation at relatively low levels of cover. Perhaps as important was that there was no evidence for a threshold amount of cover, below which there was a catastrophic decline in the number of species (Cunningham *et al.* 2014a; Cunningham *et al.* 2014b). Rather, the relationship between cover and overall bird species richness (as well as the richness of birds of conservation concern) was a gradual 'diminishing returns' curve, with the steepest changes in richness occurring with increases at low levels of cover (see Fig. 4.1). Collectively, the research suggests there is no 'magic number' for the amount of native vegetation cover that should be on a farm or in a landscape. Of course, more is better, but farms with low levels can still have significant value for biodiversity and should instead form the basis for future restoration efforts to increase levels of native vegetation cover.

The size of plantings

Size often matters in ecology, including restoration ecology. Larger plantings typically support more species, such as has been documented for birds

(Lindenmayer *et al.* 2010b) and more recently in our studies of reptiles (Michael and Lindenmayer 2018). Bigger areas typically support more species of animals and plants; this is a 'general law' in life sciences called the species–area curve. Larger areas usually support more kinds of habitats, thereby increasing the chances that an area will contain the particular kind of habitat needed by a certain species. Larger areas also support more individuals of a particular species, meaning that those species have a greater chance of persisting (rather than going locally extinct).

Although small plantings support fewer native species than large plantings, it does not mean that small plantings are without value; even plantings as small as a single tree can be provide important habitat resources (see Box 4.1). Some species of birds and reptiles will occur in small plantings, but would otherwise not persist without them. In addition, as we discuss elsewhere in this chapter, the value of small plantings can be increased if they are established near other plantings or patches of remnant vegetation or if they are intersected with other plantings. In addition, a small planting can be an anchor point around which to subsequently expand a planted area.

Figure 4.2. Relationships between the size of a planting and the occurrence of the White-plumed Honeyeater. Photo by Dave Smith.

Box 4.1. Micro-plantings

The core theme of this section of Chapter 4 has been that larger plantings are superior to small plantings. But there are cases where establishing even a single tree – a micro-planting – can be important. A classic case is where a planting is being used in an attempt to reverse the current crisis in the rapid loss of scattered paddock trees across large parts of Australia's grazing environment. Single fenced trees to prevent grazing and physical damage by livestock can help not only grow new cohorts of large old trees but also recreate the scattered spatial pattern of trees that allows them to act as stepping stones to facilitate the movement of birds and other animals throughout agricultural landscapes (Fischer and Lindenmayer 2002; Manning *et al.* 2006).

Figure 4.3. Micro-planting to establish scattered trees. Photo by Mason Crane.

The shape of plantings

Several studies have shown that wide plantings (sometimes called block plantings) support more species than narrow (or strip) plantings, a result that is particularly pronounced for birds (Lindenmayer *et al.* 2010b; Lindenmayer *et al.* 2016b). The differences in bird species richness between narrow plantings (those 25 m or less) and wide plantings (>70 m or more) is quite stark – it is ~20% lower in narrow plantings in winter (when some migratory species like the Flame Robin are present) and in spring when many species are breeding (Lindenmayer *et al.* 2016b). Reptile species richness is also higher in wide plantings.

Box 4.2. Don't give up on small and narrow plantings

Just because plantings are small in size or narrow in width does not mean they are without value. Far from it. They can still provide valuable habitat for some species. Some planting area is better than none at all. Moreover, existing planted areas can sometimes be supplemented by additional management interventions. For example, when the time comes to replace old fences around plantings, they can be widened with extra lines of trees. Very small plantings may be enlarged in a similar way. Working to maintain what you have and then seeking to improve it are core general principles of effective restoration programs.

Several factors may explain why wide plantings tend to support more species of birds and reptiles than narrow plantings. First, our recent research has indicated that wide plantings tend to have better developed shrub and understorey layers (Lindenmayer *et al.* 2018) and this, in turn, may mean such areas are better places for nesting and foraging by birds, thereby supporting more species. Second, wider plantings support more interior areas away from edges where temperatures and wind speeds can be high (Cleugh 2003). This may be an additional or alternative reason why bird species richness increases with increasing planting width. It also may explain why some species of birds and reptiles are more likely to breed successfully in wider plantings (Bond 2012). Third, levels of nest predation may be higher at the edges of plantings and therefore large plantings provide ground and shrub-nesting species with more cover and greater opportunity to successfully raise their young. Fourth, insect populations are often different at the edges of plantings relative to the interior of plantings (Ng *et al.* 2017) and this may affect the abundance of insect-feeding animals within restored areas.

The topography of the landscape

Not all parts of landscapes are created equal in terms of the values for wildlife on farms. Creeklines and areas adjacent to streams are often high in species richness, especially for birds, frogs, mammals and plants. Gullies are where water and nutrients are more abundant and available than elsewhere in the landscape. Vegetation growth and total biomass are likewise naturally higher in these places than in other parts of farming landscapes, which in turn leads to greater flowering, seed production and levels of insect populations (that are a food source for other animals). It is not surprising then that plantings established in gullies and creeklines typically support higher bird species richness than plantings on midslopes and ridges (Lindenmayer *et al.* 2010b). Plantings around farm dams also can be extremely valuable habitats for biodiversity, particularly amphibians (Hazell *et al.* 2001; Hazell *et al.* 2004; Romanowski 2009).

Figure 4.4. The Grey Fantail, a species significantly more likely to occur in wide plantings than narrow plantings. Photo by Graeme Chapman.

Hilltops are often selected for revegetation, partly because they are of less value for cropping and/or grazing, and also because they can be water recharge areas where vegetation removal had led to problems with rising water tables and secondary salinity (Stirzaker *et al.* 2002). Hilltops also can be valuable habitats for reptiles, especially when they support rocky outcrops (Michael *et al.* 2008; Michael and Lindenmayer 2018). Other groups of animals such as butterflies and some

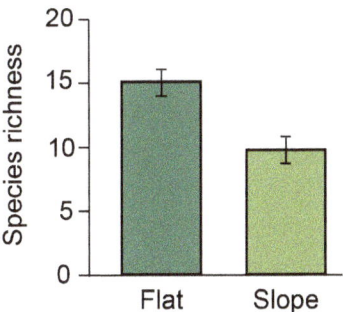

Figure 4.5. Relationships between the number of bird species in a planting and the location of the planting – flat areas and gullies versus slopes and hilltops.

migratory birds also often use the tops of hills (Michael and Lindenmayer 2018). Hilltops and ridges can therefore be good places for planting, although the spacing of trees needs to be controlled to avoid significantly altering light and radiation for temperature-sensitive animals like reptiles (Michael *et al.* 2010) (see Chapter 5).

Plantings in creeklines, around farm dams and on hilltops (especially those with rocky outcrops) all pose some challenges for management, particularly with respect to the location of fences, access to livestock and the spacing and hence density of trees. We return to the topic of planting management in Chapter 5.

Finally, plantings on north- and west-facing slopes are often more suitable for reptiles than those on (cooler and more sheltered) south- and east-facing slopes. This is because of the dependence of reptiles on warmer environments. However, not all reptiles bask directly in the sun to increase their body temperature. Some species such as the Three-toed Earless Skink spend most of their time beneath embedded logs or they move about in the humus layer of deep, loamy soils. Several of our studies have found this species responds positively to dense tree plantings (Michael *et al.* 2011) and heavily shaded forested areas (Michael *et al.* 2017).

Adjacency to other areas of native vegetation

The value of a planting for biodiversity can be strongly influenced by its adjacency to other native vegetation, including other plantings. Plantings established around paddock trees tend to have higher bird species richness than those without these structures (Lindenmayer *et al.* 2010b). There are at least two reasons for this. First, paddock trees are nearly always large old trees and are what are termed keystone structures because they have disproportionate and very positive effects on biodiversity relative to the areas they occupy (Manning *et al.* 2006). Planted areas are often dominated by young trees, and without large old trees, they lack the key resources needed by many species of animals. Second, areas with both large and small trees can increase the range of foraging and nesting niches available to be filled by different species – the likely reason why more bird species occur in plantings that also support one or more paddock trees (Lindenmayer *et al.* 2010b). There is a further advantage of establishing plantings around paddock trees because this will benefit the large old trees themselves; they tend to be in better condition than when isolated in grazing and cropping paddocks.

The value of a planting for biodiversity is often enhanced when it is located near other plantings (Lindenmayer *et al.* 2010b). This is likely because it increases the overall area available for animals to inhabit (the size issue discussed earlier in this chapter). The importance of adjacency of plantings is also underscored by work indicating that the intersections of different plantings are often characterised by higher bird species richness than elsewhere in planted areas (Lindenmayer *et al.* 2007). To further explain this concept, we compared bird species richness at the intersection of narrow plantings, along the arms of the same plantings, with isolated narrow plantings and large block-shaped plantings. We found no significant difference in bird species richness between planting intersections and large block plantings, but significantly fewer bird species were detected in isolated

Figure 4.6. Relationships between the occurrence of the Rufous Whistler in a planting and the amount of surrounding native vegetation cover. Photo by Graeme Chapman.

Figure 4.7. Intersecting plantings are often nodes of higher bird species richness relative to other parts of planted areas. Photo by Damian Michael.

Box 4.3. How plantings at one spatial scale can affect biodiversity at another spatial scale

Almost all planting is done on sites that range between half a hectare to 2 ha. It is well established that restoration at this scale can have positive effects on birds and other animals that colonise these areas (Kinross 2004; Martin *et al*. 2004; Kavanagh *et al*. 2007; Barrett *et al*. 2008; Kinross and Nicol 2008; Selwood *et al*. 2009; Lindenmayer *et al*. 2016b). However, other work indicates that these site-scale plantings can produce positive effects for bird biodiversity at a whole-of-farm scale and even a landscape scale. That is, bird species richness is higher in landscapes and on farms because of restoration efforts at a smaller (site-level) scale (Cunningham *et al*. 2014a; Cunningham *et al*. 2014b). Similarly, bird species richness on farms is significantly greater when site-level plantings are among the portfolio of vegetation assets on that farm (Cunningham *et al*. 2008). The reasons for this result remain unclear and indeed may never be determined. However, they might simply be a reflection of the value of increasing the overall amount of native vegetation cover at site, farm and landscape scales, with the increased area of vegetation resulting in increasing species richness (the species–area relationship touched on at the outset of this chapter).

Figure 4.8. The Common Brushtail Possum (top) and Common Ringtail Possum (bottom) – two species of mammals more likely to occupy patches of remnant native woodland surrounded by plantings. Common Brushtail Possum photo by Tabitha Boyer and Common Ringtail Possum photo by Damian Michael.

narrow plantings. This important finding suggests that gains in bird species can be achieved in places that have limited opportunity to establish large, wide plantings.

The value of plantings as habitat for groups of birds is elevated when they occur adjacent or near to patches of remnant woodland (Lindenmayer *et al.* 2010b). Conversely, the occupancy of remnant patches by arboreal marsupials is increased when such areas are surrounded by native vegetation cover, including plantings (Lindenmayer *et al.* 2017b). This could be because the juxtaposition of patches may increase the overall area of suitable habitat for particular species. Remnant vegetation and planted vegetation can provide the greater range of food and nesting resources required by some species, thereby increasing the chances they will occur in places where both kinds of broad vegetation types occur (Cunningham *et al.* 2008).

Although the preceding text in this section indicates that plantings support more species if they are located near other areas of native vegetation, this does not mean that highly isolated areas are without value. We have often been stunned by how even apparently remote restored areas can be colonised by many species of native woodland birds, sometimes well within 10 years of plantings being established. In fact, most species of birds do not appear to exhibit a strong negative effect of patch isolation in temperate woodland environments (Driscoll and Lindenmayer 2009). This is good news as it means that any restoration efforts by farmers and other resource managers can make a positive difference to enhancing

Box 4.4. Some places should never be planted

Planting native woodland trees should be avoided in some areas. Native grasslands is one of these kinds of places, especially grasslands that are largely intact and have not been heavily modified by ploughing, conversion to swards of exotic grasses and other forms of pasture improvement for livestock. Australia's lowland temperate native grasslands are one of the most extensively cleared and heavily modified natural ecosystems, with estimates that only 1% of it remains relatively intact. Because of their status as endangered ecosystems, native grasslands should never be targeted for replanting programs, especially tree plantings. Native grasslands can be valuable environments for livestock grazing (Simpson 1993), especially as they can recover quickly after fire and also following drought. Native grasslands are also important for some elements of the biota, and our work suggests that they can be a valuable part of the portfolio of natural assets on a farm and boost overall levels of biodiversity on farms (Cunningham *et al.* 2008). Some species of reptiles are most abundant in areas of woodland remnants and plantings where the ground cover is dominated by native grassland. These include the Striped Legless Lizard, Olive Legless Lizard, Curl Snake and Tessellated Gecko (Michael *et al.* 2004; Michael and Lindenmayer 2018).

Figure 4.9. (Clockwise from top) Striped Legless Lizard, Curl Snake and Tessellated Gecko, which are strongly associated with native grasslands as well as plantings where the ground cover is comprised primarily of native grasses. Striped Legless Lizard by Matt Clancy, other images by Damian Michael.

biodiversity in landscapes where there is only limited remaining native vegetation cover (Cunningham *et al.* 2014a; Cunningham *et al.* 2014b) and where that cover is relatively distant from other plantings or patches of remnant vegetation. Of course, not all species will respond in such heavily altered landscapes; dispersal-restricted species like the Brown Treecreeper and Squirrel Glider are examples (Walters *et al.* 1999; Cooper *et al.* 2002), especially the latter which generally cannot travel across areas where trees are more than 70 m apart (van der Ree *et al.* 2004).

Figure 4.10. A sward of native grassland – an area that should never be targeted for planting. Photo by Damian Michael.

Summary

In general plantings that are better for biodiversity will be those that:

- are large
- are block-shaped (i.e. as wide as possible)
- are adjacent to other plantings or areas of native vegetation such as remnant woodland and native grasslands
- are located in gullies
- are connected to other plantings
- support several layers of vegetation, including an understorey and native ground cover.

However, plantings that do not meet all of these criteria are not without value and can still make an important contribution to biodiversity conservation. They also might be expanded or connected at a later stage.

5

Ways to manage plantings

Plantings, much like farms, require management to succeed. This chapter explores the range of kinds of management that can be required to ensure that plantings are effective habitats for wildlife on farms. This chapter is not only informed by our scientific studies over the past few decades, but it also contains insights generated from the on-farm experiences of the landholders with whom we have worked for many years. The topics range from expected ones like controlling grazing pressure to tackling the never-ending challenge of weed and feral animal invasions. However, we also touch on management issues that are not well known or understood by some landholders and resource managers, such as the problems caused by barbed wire fencing and the negative effect of high levels of tree density in plantings established around rocky outcrops. The topics in this chapter are often those that trigger the greatest levels of discussion among farmers and, like the other parts of this book, we welcome feedback from readers on how the content and recommendations of this chapter might be improved.

Avoid clearing – of remnant woodland, natural regrowth and plantings

Land clearing has highly significant negative impacts on a wide range of native plants and animals. This is because it either directly kills them or removes enough habitat that remaining individuals often do not persist very long, especially in the case of most animals. As we have demonstrated in several places in this book, plantings provide valuable habitat for a wide range of species, as does old growth woodland and natural regrowth woodland. Clearing of all of these kinds of habitats should therefore be avoided wherever and whenever possible.

Box 5.1. You don't know what you've got til it's gone

Readers might recognise this famous line from Canadian singer Joni Mitchell. It resonates, too, with a key principle in restoration ecology. That is, it is more effective to conserve and build on what you have rather than try to recreate it once it's lost. Natural assets like patches of remnant bush and large old scattered paddock trees are almost impossible to replace, even by the best designed and managed plantings. As an example, nest boxes can be useful (if they are designed well, carefully maintained and regularly replaced), but they can never provide all the other services supplied by large old hollow-bearing trees, such as fruit, seeds, flowers, bark where insects live, perches, and carbon storage. In the case of such natural assets as these on farms, we do know the value of what we've got and should strive to ensure it does not disappear.

In some cases, plantings have been established in an attempt to compensate for the clearing of old growth woodland. This is a strategy called biodiversity offsetting and it is widespread in Australia and overseas. There is no doubt that establishing plantings is a good thing. However, plantings are not effective offsets for the loss of patches of remnant woodland and scattered paddock trees. There are good reasons for this. First, plantings, regrowth woodland and old growth woodland are markedly different kinds of habitat that are used by quite different assemblages of species of birds (Lindenmayer *et al.* 2012a) and reptiles (Michael *et al.* 2010). Less than half of the reptile species found in the South West Slopes region of New South Wales occur in plantings. Similarly, about three times as many species of birds inhabit old growth woodlands than plantings, which means plantings cannot replace the habitat removed by clearing. Second, it can take a very long time for plantings to be colonised by some species, particularly cavity-dependent animals like possums and gliders (Lindenmayer *et al.* 2017b). Third, plantings (and regrowth in particular) perform necessary roles as refugia for small-bodied birds and migratory birds during periods of drought and extreme drought (Lindenmayer *et al.*, unpublished data) – conditions that will become increasingly prevalent with climate change (Steffen *et al.* 2009).

Control the intensity of livestock grazing

The issue of grazing impacts on biodiversity and ways to mitigate them is an immense topic and a thorough treatment could fill an entire volume. Some of the science associated with grazing impacts is only very briefly considered here, and only as it relates to grazing's direct and indirect effects on biodiversity in plantings. Our long-term studies suggest that prolonged and intense grazing (i.e. high stocking rates) of plantings reduces bird species richness (Lindenmayer *et al.*

Figure 5.1. Olive Legless Lizard, which is a native species of reptile that occupies plantings, especially those that have not been grazed or where grazing intensity and frequency is limited. Photo by Mason Crane.

2010b) and also reptile species richness (Lindenmayer *et al.* 2018). It also affects a range of individual bird and reptile species such as the Olive Legless Lizard, which prefer to shelter in large grass tussocks.

Livestock grazing can have direct and indirect effects on both the condition of the vegetation within plantings and on the biota that occupies these areas (Kay *et al.* 2017). In general, high-intensity set stocking grazing (where animals remain continuously in the same paddock, often in high numbers) has the most substantial negative impacts on biodiversity through altering the ground layer, damaging the shrub layer, and impairing the regeneration of understorey and overstorey trees (Martin and McIntye 2007; Sato *et al.* 2016). The leaf litter layer can be particularly altered by grazing and this, in turn, influences the species richness of birds and reptiles as well as several individual bird and reptile species (Lindenmayer *et al.* 2018) (Fig. 5.2). Grazing also modifies the suitability of habitats for invertebrates such as ants and beetles (Barton *et al.* 2016) which are, in turn, sources of food for animals such as birds and reptiles. For ground-nesting bird species like the Rufous Songlark and the Bush Stone-curlew, the effects of livestock grazing can result in the direct trampling of their nests.

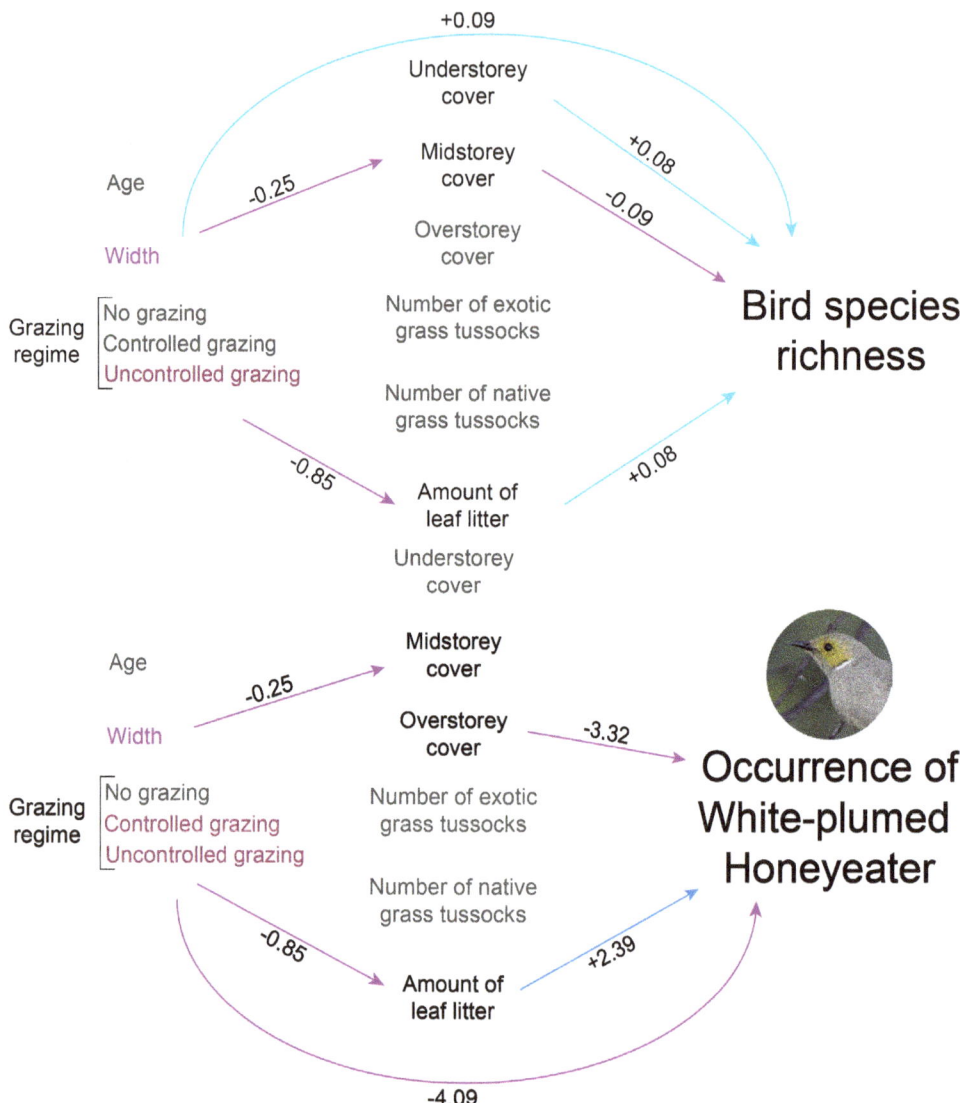

Figure 5.2. Path analysis diagrams showing how grazing and other characteristics of plantings can influence biodiversity (in this case bird species richness and the occurrence of the White-plumed Honeyeater) through changing vegetation condition (e.g. the litter layer), which then affects the suitability of habitats for wildlife (redrawn from Lindenmayer *et al.* 2018).

Intensive grazing of established native vegetation around watercourses and farm dams can reduce or eliminate riparian area plants as well as aquatic native vegetation, leading to significantly impaired habitat quality for frogs (Hazell *et al.* 2001; Hazell *et al.* 2004; Romanowski 2009). Direct stock access to dams and water courses can lead to fouling of waterbodies with subsequent negative impacts on water quality for livestock (and livestock health).

Figure 5.3. Rufous Songlark – a ground-nesting native bird species that is significantly more likely to occur in plantings which have not been grazed. Photo by Graeme Chapman.

Beyond the negative effects of intensive set stocking, other kinds of grazing regimes may have less substantial impacts on biodiversity. For example, our research to date suggests that crash grazing in plantings, especially in winter when ground cover plants are not flowering may have relatively limited effects, although further monitoring is needed to confirm the results we have gathered to date.

Fencing is one of the primary ways in which grazing pressure can be regulated both over time and spatially across a farm. Appropriate fences and gates can provide controlled access to water points within plantings (see Fig. 5.4), which will help maintain vegetation condition and the quality of water (which benefits livestock) (Romanowski 2009).

Consider the kind and condition of fences around plantings

It is a truism to state that fences on farms are critical infrastructure. They are invaluable for controlling livestock movement and grazing pressure in planted areas. Fences are also used as perching sites by many species of birds, including those of conservation concern like the Flame Robin and Southern Whiteface, or as basking sites for reptiles such as sun-loving Bearded Dragons and Wall Skinks. However, some types of fencing can have significant negative effects on wildlife. Barbed wire is

Figure 5.4. Vince Ryan on his farm showing the gates and fences he installed around creeks and dams. This has helped control grazing around waterpoints and led to improved condition of planted vegetation. Fencing also prevents stock fouling dams and thereby enhances water quality, leading to better condition and weight gain of livestock. Photo by Mason Crane.

a case in point. Many thousands of native animals get caught on and killed by barbed wire fences every year. Some are common like the Australian Magpie and the Eastern Grey Kangaroo, and losses of many individuals will make little long-term difference to populations. Others are rare or threatened and deaths on barbed fences can ultimately have non-trivial negative effects on wildlife populations. Examples of such species of conservation concern include the Squirrel Glider, Greater Glider, Grey-headed Flying Fox and the Little Red Flying Fox (Fig. 5.5).

Box 5.2. Mature trees, biodiversity and grazing in older plantings

Some farmers think that when the trees are big enough to handle grazing the job is done and fences can be removed. However, they may not realise that many of the biodiversity gains from plantings can quickly be reversed by intensive grazing. This can have negative impacts on tree health, and can result in the death of trees. These problems are especially prominent in small and narrow plantings, where there can be high concentrations of livestock seeking shelter. However, controlled grazing is not always destructive. Some tree plantings become vulnerable to exotic grasses and other weed species that can be prevented through the implementation of strategically controlled grazing regimes.

Figure 5.5. This bat is an example of the dangers of barbed wire fences for wildlife. Photo by Damian Michael.

Electrified fencing or replacing barbed wire with plain wire on the top rung of a fence are solutions to this problem. Many landholders with whom we have worked have done this successfully and without compromising stock access to key parts of their properties. These changes can be particularly appropriate when it comes time to replace fences in parts of farms and around plantings. As an immediate measure to protect gliders, polypipe can be cut to length and fitted over the top two to three strands of barbed wire fencing in areas that are regularly used as a flight or gliding path by Squirrel Gliders (Fig. 5.6). This might be between a regularly used den tree and a frequently used tree planting. Landholders who have implemented this preventative method have reported a marked reduction in gliders becoming entangled in fencing wire.

The location of fences around plantings in gullies and creeklines needs particular thought as periodic flood events can wash expensive fencing materials downstream if they are set too close to the instream channel. Setting fences back from where water flows in floods can overcome this problem, as can avoiding constructing fences across channels. Another solution is to use a flood-proof fencing design or flood gates. Designs include drop-down fences that give way under the pressure of flood waters to lie flat on the ground, and can be re-erected after the flow has passed (Bell and Priestley 1998; Romanowski 2009).

Figure 5.6. Polypipe can be used to protect gliders from barbed wire fences. Photo by Damian Michael.

Box 5.3. No fences and drowned sheep

One of our landholders recounted the underlying reason that triggered him to fence his creeklines. His sheep liked to graze in deeply eroded gullies and ahead of an impending storm he moved them upslope. A lack of fences resulted in them returning to their favourite haunt in the gully. The subsequent flash flood drowned the entire flock and washed them into the neighbouring property. All of the creeklines on the farm are now fenced, have access waterpoints for stock, and the plantings within these fenced areas look spectacular.

The condition of fences is one of the biggest issues associated with managing grazing pressure (Spooner *et al.* 2002). Key changes in vegetation condition (like the amount of natural regeneration) are typically associated with the amount of time elapsed since fences were established (Spooner and Briggs 2008). However, the value of fences in modifying grazing regimes can be rapidly eroded if fences become degraded and are not replaced, as is occurring in some areas restored under replanting programs completed 10–20 years ago under initiatives like the Natural Heritage Trust. Replacing fences is an important part of managing the condition of planted woodland. It also provides a chance to alter management such as by widening narrow block plantings to make them more effective for wildlife (see Chapter 4).

Think about the density (tree spacing) of plantings

Temperate woodlands are inherently open environments; considerable amounts of light reach the ground and hence spacing between trees is important (Lindenmayer *et al.* 2005). Light and therefore temperature regimes can be important for some ground-dwelling plants and also many species of reptiles. Densely stocked plantings can alter these fundamentally important micro-environments with negative impacts on some species like skinks and small snakes. Establishing plantings with a high density of trees can also result in increased mortality of some individual stems as they lose the competition for key resources like water. The density of plantings needs careful thought in areas that might otherwise be important habitats for reptiles, such as on hilltops and rocky areas (see Chapter 3).

There is no general recipe for the density of trees for establishing plantings as it can vary in response to factors like the species of plants being established, rainfall, elevation, topography, soil capability and productivity of the land. It is best to seek advice from professional restoration practitioners like staff from Catchment Management agencies, Local Land Services, Landcare and Greening Australia. However, a general rule of thumb would be to design revegetation programs based on the existing ecological vegetation communities in the landscape. Thus, in high rainfall areas, forest species can be used and in low rainfall areas, low-growing shrubs and subshrubs such as saltbush species will grow best.

Figure 5.7. The Tree Crevice Skink, a species negatively affected by densely spaced tree plantings that change micro-climatic conditions around its rocky outcrop habitats. Photo by Damian Michael.

Control weeds and feral animals

Invasive plants and animals are a constant problem for almost all landowners. Most agricultural landscapes are extensively disturbed and the majority of invasive species do better in such modified environments than in more intact locations. Plantings are no different in this regard and need to be managed. Foxes need to be shot or poisoned. Rabbit control requires not only these measures, but also the excavation or 'ripping' of warrens (although this should only be done at an appropriate time of the year when there will not be adverse effects on other animals such as the Carpet Python) (Michael *et al.* 2010) (see Box 5.2). Weeds such as St John's Wort and Blackberry must be sprayed.

A key lesson in almost all invasive species control programs is that they are most ecologically effective, and generally most cost-effective, when managers intervene early – that is, before invasive species are widespread. Of course, this is not possible in all circumstances, particularly with some already very widespread pests. In some cases, constant and repeated management is needed to contain invasive species problems and to keep a weed species restricted before it becomes widespread.

Box 5.4. Management for an iconic Australian reptile

The Carpet Python is well regarded by landholders for its role in controlling pest animals such as mice, rats and rabbits. It was once a common practice for landholders to capture pythons and relocate them to grain and hay sheds to help control rodents. Pythons are also well adapted to climbing trees where they raid the nests of cockatoos and prey on possums. Unfortunately, this remarkably patterned snake has declined considerably in north-eastern Victoria and southern New South Wales over the past several decades. The main causes of its decline are attributed to habitat loss, habitat degradation, poaching in some areas, and a loss of medium-bodied native prey. Native mammals such as Bandicoots, Bilbies, Bettongs and Bush Rats have long since disappeared from agricultural landscapes in south-eastern Australia.

Recent studies using radio-telemetry to understand how these iconic snakes use farming landscapes has found that many individuals spend large amounts of time during the spring and summer occupying rabbit warrens (Heard *et al.* 2004). Here, they have access to a ready supply of food and stable warren temperatures help them regulate their own body temperature. During late summer and autumn, Carpet Pythons tend to take to the trees to capture the last rays of sun before they spend the cold months sheltering within deep crevices in rocky outcrops (Michael and Lindenmayer 2008). A major observation of this work is that pythons use different habitats throughout the year, so by ripping warrens during the cooler months, landholders are less likely to inadvertently kill or injure a sheltering python (or other reptile) (Michael and Lindenmayer 2018).

Figure 5.8. Carpet Python. Photo by Damian Michael.

Box 5.5. The need to tackle the 'big 3'

The Australian continent suffers greater negative effects of feral animals than almost any other part of the world. Many extinctions of native animals are a result of introduced predators like the Fox and Feral Cat (Woinarski *et al.* 2015). Rabbits are often a key prey item of these exotic predators. Large populations of rabbits often result in high populations of a predator that also then kills many native animals. Reducing populations of just one of them (e.g. the Red Fox) can be problematic because it 'releases' populations of the others, particularly cats. A key task in pest control is to limit populations of all three introduced animals – the prey and the predators at the same time.

In the case of wide-ranging pest animals like the Red Fox, coordinated baiting programs are essential to reduce populations across reasonably large areas. This means groups of farmers working together to deliver a coordinated feral predator control program over large areas and for prolonged periods. Failure to do this can result in baiting programs being ineffective, in part because of what are sometimes termed 'vacuum effects' – areas with depleted numbers of pest animals will remain empty only very briefly as other individuals are drawn in to occupy the temporarily emptied territories. There are some excellent examples of highly successful fox baiting programs. The Bounceback initiative in the central north of South Australia was based on coordinated baiting and shooting across many properties and there fox control has resulted not only in the recovery of the iconic Yellow-footed Rock-wallaby, but also significantly elevated rates of lambing success – a win for farmers and a win for conservation.

Box 5.6. It's not just the feral animals that need to be controlled

Sometimes populations of native animals can reach a size where they need to be controlled not only because of damage to agricultural production but also because of the damage they can do to the natural environment. For example, there is now overwhelming evidence that populations of herbivores such as the Eastern Grey Kangaroo can reach a size where they have significant negative impacts on birds, small mammals, reptiles, beetles, spiders and native plants (Barton *et al.* 2011; Howland *et al.* 2014; Foster *et al.* 2015; Foster *et al.* 2016). Large populations of kangaroos develop because humans have created ideal conditions for them – water (through increasing the number of farm dams and watering points) and grass. Moreover, hunting by Aboriginal Australians reduced populations of kangaroos as did predators like the Dingo (and probably the Tasmanian Tiger and Marsupial Lion before that), which are now either substantially reduced or extinct. A different kind of 'predation' through culling may be needed to keep populations of kangaroos in check. Killing native animals is never a welcome activity, but it can be an essential part of farm management as well as biodiversity conservation (although see Box 5.7).

Figure 5.9. A large mob of kangaroos. High numbers of herbivores such as the Eastern Grey Kangaroo can have negative impacts on not only the condition of native vegetation but also on native mammals, birds and invertebrates (such as beetles, ants and spiders). Photo by Bob Bowker.

Box 5.7. When shooting does not work – and a longer term planting strategy is better

In several parts of this book we describe the negative impacts of the hyper-aggressive Noisy Miner. The species has now been listed as a key threatening process in the temperate woodlands because of its effects on small woodland birds, many of which are of conservation concern. One strategy suggested to tackle the Noisy Miner problem has been to cull them. A PhD program by Richard Beggs at The Australian National University is experimentally testing if this is an ecologically effective and cost-effective solution. Noisy Miners have been humanely removed from woodland patches and the effects closely monitored and compared against similar patches where no culling has taken place. The overwhelming results to date are that Noisy Miner control is very short-lived (typically for only a few weeks) before the patches are recolonised by new birds. Culling is also expensive. The preliminary conclusions are that it is better to take a longer term strategic view and establish plantings (which miners do not like) as the most ecologically effective and cost-effective solution to this over-abundant native species.

Maintain dead trees and logs

Large old trees and large pieces of fallen timber are critical habitats for many native species. For example, almost all species of possums and gliders and parrots nest only in cavities in large old trees, structures that can take decades if not centuries to develop. Logs are similarly important habitats for numerous species. For example, the presence of large old trees and fallen logs in plantings can significantly elevate the number of bird species that occur in these habitats (Lindenmayer *et al.* 2010b).

Box 5.8. Resist the temptation to tidy up

Many landowners like the look of a tidy farm, preferring to remove dead trees, fallen branches and fallen trees. However, tidying up this debris can have devastating effects on wildlife, as it serves as habitat for a range of native plants and animals. Removing fallen trees from paddocks and fencelines (such as by burning them) is understandable from a practical management perspective, especially in areas targeted for intensive cropping. However, the value of fallen timber for wildlife can be maintained in creative ways like relocating it into plantings where we know it will have important biodiversity value (Lindenmayer *et al.* 2010b) and provide habitat for animals (such as the Carpet Python) that can play key prey control roles on farms (see Box 5.4).

Don't remove mistletoe

Mistletoe is a semi-parasitic plant that draws nutrients from the host trees. It has a bad reputation, as it can potentially kill the host tree when infestations become very large. However, mistletoe is also an important source of food for a range of bird species as well as several mammals and invertebrates. The value of mistletoe for wildlife is clearly demonstrated when experiments deliberately remove it and the species richness of birds declines significantly (Watson 2001). In the case of restored areas, bird species richness is significantly higher in plantings where mistletoe is prevalent (Lindenmayer *et al.* 2010b).

Add nest boxes – but use the right design and ensure they are well maintained

Large old trees provide nesting habitat for hollow-dependent animals (Manning *et al.* 2006). Populations of large old trees (often termed paddock trees) are in serious decline in many grazing and cropping landscapes not only in Australia (Fischer *et al.* 2009) but worldwide (Lindenmayer and Laurance 2016).

Large old trees are one of the features of old growth woodland that are often rare in, or even totally missing from, plantings. This limits the use of plantings by some cavity-dependent animals, although animals like the Squirrel Glider may use them for foraging even if they cannot nest or den there (Crane *et al.* 2014). The prolonged period that it takes for most woodland trees to develop cavities means that it will be many decades or even a century or more before the trees used to establish plantings will eventually develop suitable hollows.

Establishing nest boxes is one management approach often used in an attempt to tackle the problem of a paucity of hollow trees in plantings. We have completed several studies to quantify the effectiveness of nest boxes as a source of cavities for hollow-dependent animals in temperate woodlands (Lindenmayer *et al.* 2016a; Lindenmayer *et al.* 2017a). Nest boxes are used by some hollow-dependent fauna but principally by already common species (including exotic pest species) and generally not animals of conservation concern (Lindenmayer *et al.* 2016a). Nevertheless, nest boxes can be valuable for some species and enable them to colonise plantings when they otherwise would be unable to do so (Lindenmayer *et al.* 2016a). They can be most effective when plantings are connected to patches of remnant native woodland, which appear to be sources of young dispersing into previously unoccupied restored areas (Lindenmayer *et al.* 2016a).

Our collective studies suggest there is a need for careful attention to the design of nest boxes so that they are more likely to remain functional for long periods of time and meet the requirements of particular target species. This includes animals that are threatened and that can have specialised nesting requirements. Low levels of occupancy by key species of conservation concern, coupled with high rates of attrition (and hence the need for frequent replenishment) indicate that nest boxes are a very poor substitute for large old trees in grazing and cropping landscapes. Better protection of existing paddock trees is critical as they are long-lived and difficult to replace. Maintenance of populations of these keystone structures is an essential part of conservation actions to enhance the protection of biodiversity in agricultural landscapes.

A key lesson from our nest box studies is that nest boxes need to be well designed (see Goldingay *et al.* 2015) to be effective and not have perverse effects like increasing populations of pest species. They will also require regular maintenance to be useful to wildlife. Appropriate design is especially important if the aim is to recover populations of species of conservation concern. Management agencies like Local Land Services and wildlife authorities will often have technical information on the kinds of designs of nest boxes that are likely to be most effective for particular target species.

Avoid or limit some kinds of management

Some kinds of management are highly damaging and should be only rarely practised or excluded entirely from the management toolkit. The opening sections of this chapter highlighted the damage to biodiversity associated with clearing native vegetation. But there are other kinds of habitat for which some kinds of management can be highly detrimental. For example, bushrock removal is a key threatening process and has major long-term negative effects on reptile habitat, as many reptiles do not survive without access to these environments. A disturbing practice that is emerging in the agricultural sector is the use of heavy rock-crushing machinery (called the 'Reefinator') to pulverise volcanic rock on the western plains of Victoria as well as parts of Western Australia and South Australia (Michael and Lindenmayer 2018). Expanding intensive agricultural practices into places that were not previously subject to high-impact management practices contrasts markedly with conservation objectives and the principles of land sharing, whereby landholders strive to integrate biodiversity into production landscapes. Other practices like firewood removal, especially when it is targeted at large old dead trees, are similarly damaging, mainly for the array of species known to nest and/or den in the cavities in these trees (such as the Superb Parrot and the Squirrel Glider). For example, almost 10% of the dead hollow-bearing trees used by the Squirrel Glider were deliberately cut for firewood in just one six-month period (Crane *et al.* 2008). Controlling these activities means controlling who comes onto a property. Landholders interested in sustainable farm management should therefore restrict access to bushrock and firewood contractors coming onto their land.

Figure 5.10. Nest boxes with different designs for different species of mammals and birds (clockwise from top left, rosella box, kookaburra box, bat box, phascogale box, glider box with additional bat entrance). Photos by Mason Crane.

Finally, there is increasing evidence that some traditional farming practices like intensive and prolonged set stock grazing within plantings, as well as the extensive use of chemicals near plantings, can have profound detrimental effects on wildlife on farms. We suggest that all of these practices should be limited or avoided wherever and whenever possible.

Figure 5.11. Large old dead hollow-bearing tree used by the Squirrel Glider, the kind of tree often targeted for firewood harvesting. Photo by Mason Crane.

Plan for fire

Australia is the most fire-prone continent on Earth and, as all farmers know, fires occur in agricultural landscapes across the nation. Fire can have a significant effect on farm wildlife from accelerating the loss of large old trees (Crane *et al.* 2016), including in plantings, to destroying plantings outright, especially if the trees and shrubs are young and have not matured to a stage where they can produce viable crops of seeds. However, many eucalypt species do survive fire and readily resprout after being burned. In addition, fire can break the seed dormancy of some plant species, thereby promoting germination. This can enable at least some of the species that comprise many plantings to recover after fire.

Some farmers mistakenly believe that adding plantings to their farms will increase the risk of wildfire. There is no evidence to support this (Jenkins *et al.* 2016). However, work in parts of Victoria following the catastrophic 2009 Black Saturday wildfires suggests there are some important steps that can be taken to limit the extent of property damage and risk to humans from fire (Gibbons *et al.* 2012). Perhaps the most important recommendation is to ensure that plantings and other areas of native vegetation are established more than 30 m away from key infrastructure like homesteads, sheds and other buildings (Gibbons *et al.* 2012). In addition, it is important to put gates at regular intervals within plantings to allow rapid access to adjacent paddocks to facilitate fire suppression.

Prescribed burning will sometimes be an important part of farm management. There are some strategies that can be employed to limit the impact of burning on biodiversity and other environmental assets. First, avoid burning near large old trees as they are readily destroyed by fires, including very low-intensity and high-severity burns (Holland *et al.* 2017). Second, try not to burn all of an area at once, leaving some unburned places to provide refuges for some species and escape routes for others. Third, where possible, try to ensure that burned areas are rested

---

Box 5.9. Fire and woodlands

Australian woodlands have a long history of fire – potentially dating back millions of years. The frequency, intensity and seasonality of fires have changed profoundly since European settlement of the broader temperate woodland belt, particularly as increasing amounts of infrastructure such as homesteads, sheds and fences were established. There is a role for more fire management in some natural areas, such as travelling stock reserves and other woodland-dominated public land. However, there is generally a need to protect plantings from fire as a general rule, as it remains unclear where the trees in such places will persist after a fire and/or whether natural regeneration might occur. This is particularly the case where trees are young and may not have reached an age where they produce viable seed.

Figure 5.12. Try to avoid prescribed burning that ignites and then destroys large old trees. Photo by Mason Crane.

for a prolonged period (typically 5–10 years) before they are reburned to allow some fire-sensitive species to recover. Fourth, try to avoid prescribed burns becoming high-intensity wildfires as these can kill many animals. This means thinking carefully about the right temperature, wind and moisture conditions before prescribed burning is conducted. In many places, only a small window of opportunity exists to safely conduct a prescribed burn that is easily controllable and that minimises negative ecological effects.

Other kinds of management

Early in the history of vegetation restoration on farms, the primary focus was often on planting only overstorey trees. Very few had understorey shrubs included in the mix of species being established. The vegetation structure in such plantings can therefore be quite simplified and habitat suitability for some species is less than ideal compared to what might be available in the presence of both an overstorey and an understorey. We have implemented a small experiment over the past five years in an attempt to rectify this problem in a set of our long-term remnant

Figure 5.13. One of a dozen remnant woodland sites dominated by overstorey trees and where an understorey replanting program was conducted. Photo by Mason Crane.

woodland sites. So far, this work has indicated that such efforts can be successful in establishing additional layers of vegetation cover. The response of birds and reptiles to this management intervention have yet to be quantified, but monitoring of the array of sites in the program is continuing.

Do interventions work?

Resource managers have a range of management options available to them to improve vegetation condition, biodiversity conservation, and other environmental outcomes. These include (among others) fencing to control the timing and intensity of livestock grazing, active replanting of vegetation cover, and weed removal. A critical question is: do these management interventions work? This simple question is rarely asked and even less commonly answered. Yet it is important to know whether interventions are effective. We sought to answer this question in a study of a farm incentive scheme in the western Murray region of southern New South Wales (Lindenmayer *et al.* 2012b). This incentive scheme paid farmers to conduct management actions in a range of woodland vegetation

---

Box 5.10. Not everything works

This book contains descriptions of many kinds of management to enhance native vegetation cover, especially tree plantings. But not everything that has been tried is successful. For example, we have tried repeatedly to establish swards of native Kangaroo Grass (*Themeda australis*) as part of restoration and replanting programs. We have gathered Kangaroo Grass hay and spread it on paddocks and monitored this intervention. The results have sometimes been successful but underwhelming at other times, whereas our colleagues have been successful in establishing swards of this species of native grass (e.g. Prober *et al.* 2005; Smallbone *et al.* 2007). It is not clear why our efforts have sometimes failed, although it is possibly associated with burning before attempts to re-establish swards of grasses. Much can be learned from failed interventions (often more than ones that are successful). In this case, our results to date highlight the importance of ensuring that existing areas of native grasslands are not lost or degraded given how hard they can be to re-establish.

---

communities. Our study into this scheme showed that, relative to reference sites in nearby conventional agricultural production areas, there was a significant improvement in vegetation condition and cover on farms where the incentive-based intervention to enhance conservation had occurred. In addition, there was a change in the bird assemblage towards smaller-bodied, insectivorous woodland bird species, including some species of conservation concern (Lindenmayer *et al.* 2012b). This study provided evidence that management interventions can and do have a positive influence on vegetation cover and on some groups of animals like birds (Lindenmayer *et al.* 2012b).

Summary

In general, the best managed plantings are those where:

- fences are maintained
- livestock grazing is limited or excluded
- mistletoe, rocks and logs are not removed (and indeed sometimes supplemented, such as by adding fallen timber from surrounding cropping paddocks where it can impede farming machinery)
- an understorey is present
- weeds, exotic animals and over-abundant herbivores are controlled
- gates and other access points are established to assist in fire management.

There are therefore many ways plantings can be managed so that their value for wildlife is maximised. Indeed, often relatively minor and inexpensive management actions can have large benefits (such as leaving dead trees in plantings). Many

Box 5.11. Management of natural regeneration versus plantings

The primary focus of this book has been on deliberately planted areas of native vegetation. As outlined elsewhere in this book, natural regeneration can sometimes be a far cheaper way to revegetate farming landscapes. Should these areas be managed differently from plantings? The answer to this question is usually no. Similar to plantings, the management of natural regeneration will often entail limiting the amount and duration of high-intensity set stocking grazing by livestock (Dorrough and Moxham 2005; Fischer *et al.* 2009), and actively controlling weeds and pest animals, including feral predators like foxes. Fire management might be important, and frequent burning (such as the annual application of fire) should be replaced with burning over longer intervals (e.g. 10 or more years) to ensure plants have the opportunity to establish and recover. Importantly, and as we have emphasised elsewhere in this book, natural regeneration and plantings are different habitats for wildlife on farms. Wherever possible, both should not be cleared.

other types of management can have broader benefits on a farm such as the control of feral animals and invasive plants. Funding is often available to assist farmers to not only establish plantings, but also to conduct management programs that maintain or improve the conservation and other values of restored areas on a farm.

Figure 5.14. An example of a fence in need of repair. Good fencing is key to the effective management of planted woodland. Photo by Esther Beaton.

6

How do plantings change over time?

Nothing stays the same in ecology. This is especially true of planted vegetation. That is why long-term work and monitoring are essential to quantify changes in vegetation and biodiversity. In this chapter, we briefly outline some of the changes in vegetation that have occurred over time. We also discuss some of the changes in biodiversity that have occurred in plantings over time.

Changes in the vegetation structure of plantings

Over the past two decades we have carefully measured and remeasured many of the key attributes of the vegetation that characterise plantings. These measurements include tree height, the amount of cover of the overstorey, understorey and ground layers, and the amount of bare earth, the litter layer, and the extent of lichen cover (a surrogate for soil nutrient levels and soil disturbance). We found strong evidence for increases in values for many of these measures – overstorey trees grew taller, as did the understorey, the amount of overstorey cover increased, and there was an increase in the number of stems over time. By contrast, the amount of ground that was bare earth declined over time. There were significant positive changes in many of these measures over time. For example, canopy height increased by 0.38 m per year on average, compared to a mean height of ~10 m. The increases in these measures were approximately linear over the roughly 20-year range of our data, except for canopy depth, which plateaued after ~15 years (Lindenmayer *et al.* 2016b). Some of these findings may seem trivial (e.g. that trees get taller as they age), but documenting them is valuable for

Box 6.1. Take photos

Plantings change, sometimes remarkably quickly, and it can be easy to forget how things used to be. Taking photos can be useful to illustrate the changes that can occur with revegetation efforts. This can be important in highlighting the difference that farmers, Landcare groups and staff from natural management agencies can make through their restoration efforts. Photos taken from specified locations repeatedly over time can be a useful method for assisting monitoring in vegetation cover on farms (see Chapter 7).

understanding how particular animal species might respond to plantings over time. This kind of information is also crucial for continuously enhancing the quality of farm plans, enhancing the effectiveness of plantings, and better integrating agricultural production with biodiversity conservation (see below and also Chapter 7).

The data presented in Table 6.1 describes how the attributes of individual plantings can change over time. However, there are also changes in landscapes that result from many concerted restoration efforts in many places. For example, the amount of native vegetation cover in the South West Slopes of New South Wales has increased by more than 3.4% over the past 10 years. Approximately half of this has been the result of active restoration, and the remainder natural regeneration (regrowth) following a reduction in grazing pressure by domestic livestock (Sato et al. 2016). These changes in the amount of native vegetation cover in landscapes can, in turn, have positive impacts on the species richness of both birds and reptiles, and also appear to be important for native mammals such as possums (Cunningham et al. 2007; Cunningham et al. 2014a; Cunningham et al. 2014b; Lindenmayer et al. 2017b).

Table 6.1. Vegetation measurements investigated for changes since time since planting.

The estimates are the average increases per year over the 30-year range (excluding the outlying oldest site); the counts and percentages were analysed on the log scale, so estimates are average percentage increases per year, relative to the previous year (from Lindenmayer et al. 2016b).

Measurement	Mean and s.d.	Estimate and 95% CI
Canopy height	9.9 m (4.1)	0.4 m (0.30, 0.46)
Canopy depth	7.7 m (3.3)	0.1 m (0.06, 0.22)
No. stems/ha	935 (993)	2.9% (−0.5, 5.3)
% Overstorey cover	16 (22)	17% (11, 24)
% Midstorey cover	26 (25)	0.0% (−3.7, 3.7)
% Understorey cover	3.0 (4.1)	−0.3% (−4.6, 4.2)
% Bare earth	13 (17)	−2.1% (−6.3, 2.3)

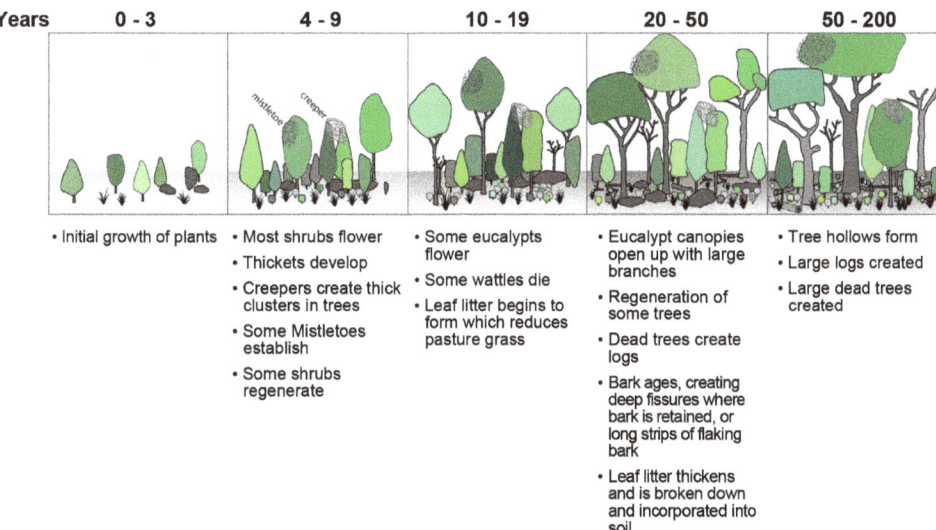

Years	0 - 3	4 - 9	10 - 19	20 - 50	50 - 200

- Initial growth of plants

- Most shrubs flower
- Thickets develop
- Creepers create thick clusters in trees
- Some Mistletoes establish
- Some shrubs regenerate

- Some eucalypts flower
- Some wattles die
- Leaf litter begins to form which reduces pasture grass

- Eucalypt canopies open up with large branches
- Regeneration of some trees
- Dead trees create logs
- Bark ages, creating deep fissures where bark is retained, or long strips of flaking bark
- Leaf litter thickens and is broken down and incorporated into soil

- Tree hollows form
- Large logs created
- Large dead trees created

Figure 6.1. Diagram showing some of the general changes over time in the structure of a planting (redrawn from Munro and Lindenmayer 2011).

Figure 6.2. Taking measurements of the structure of the vegetation in plantings. Photo by Sachiko Okada.

The changes in vegetation discussed in the preceding paragraphs of this section are typically gradual ones reflecting 'slow' processes like tree growth and improvement in vegetation condition. There are also, however, episodic processes that can have a marked impact on vegetation cover and structure. Fire is an example. Fires can very rapidly burn extensive areas, including plantings. Fires also can destroy large proportions of populations of large old trees, including those surrounded by plantings (Crane *et al.* 2016).

Some things that are missing

Although trees and other plants are growing and the amount of cover in some vegetation layers is increasing, some key elements of the vegetation in plantings can still be missing, even after many decades. One of these is large old trees as these often take a century or more to form hollows suitable for wildlife (Vesk *et al.* 2008; Lindenmayer and Laurance 2016). The absence of these kinds of trees can be mitigated at the time of planting establishment by locating plantings around existing large old trees (see Chapter 5). An alternative is to erect nest boxes within plantings and this has met with some success (Lindenmayer *et al.* 2016a; see Chapter 5).

Another key attribute of plantings that can be missing is an understorey or shrub layer. This can occur for several reasons, but it is often a relatively obvious one – only overstorey trees were initially established. Indeed, in the early years of restoration programs, some practitioners did not recognise the need for other layers of vegetation structure, electing to focus solely on trees such as eucalypts, Acacias and She-oaks. These problems are increasingly being addressed in new generations of restoration plantings – a testament to our ability to learn and

---

Box 6.2. Why do hollows take so long to develop?

Tree hollows are a crucial resource for more than 300 species of Australian vertebrates. This nation has by far the greatest proportion of cavity-dependent vertebrates of any continent. Given that so many animals depend on them, it is puzzling that hollows in Australian trees take so long to develop – often hundreds of years (Gibbons and Lindenmayer 2002). One of the reasons why hollows take such a long time to form here is Australia's lack of animals like woodpeckers that excavate cavities. Here we rely on the long, slow processes of decay mediated by termites, fungi and bacteria to eventually create the hollows that are fundamental for providing shelter and nesting sites for myriads of species. The prolonged time it takes to develop cavities in Australian trees underscores the extensive impacts that can arise when such trees are cut down and the protracted time that must elapse until new trees with hollows can be recruited (Lindenmayer *et al.* 2017a).

Box 6.3. Fungi are important

Australian trees are notoriously bad at extracting nutrients from the already nutrient poor soil. To overcome this deficiency, the roots of many tree species have developed mycorrhizal associations with fungi. Specifically, the underground (mycorrhizal) fungal mycelia penetrate the soil where they connect with the root systems of host plant species. Here they form a symbiotic relationship, the tree providing carbohydrates and in return the fungi providing various inorganic nutrients and trace elements. Healthy ecosystems are often a product of complex fungi–plant associations and in intensively farmed landscapes, soils are often depleted of beneficial microbes and fungal mycelia. Improved crop growth and nutrient levels can be achieved through improving the health of soils, a practice which is not new and gained popularity in the old soil conservation days. The concept of improving soil conditions could also be transferred to newly restored areas, whereby fungi are transplanted along with various vegetation layers to improve the overall success and health of the tree planting.

adaptively manage agricultural landscapes. In existing overstorey-only plantings, it may be possible to do an underplanting in which shrubs and other trees are established. Box 3.2 in Chapter 3 describes a study by researchers at The Australian National University where this was done. Another key component missing from tree plantings are fungi (see Box 6.3), and while we are yet to quantify patterns of fungi in agricultural landscapes, field observations indicate tree plantings support far fewer fungi species than do undisturbed woodlands.

How do populations of animals in plantings change over time?

This section describes some of the long-term changes in animal populations in plantings. It is important to recognise that some of these changes are associated with changes in environmental factors like weather or grazing pressure, whereas others are linked with the attributes of plantings like their age, width or size. For example, as plantings age, some of their attributes change (Table 6.1) and this can have an important influence on the species which occur in these places, such as through affecting the suitability of places to forage or to nest (Lindenmayer *et al.* 2016b). Where this is the case, we have tried to provide a simple description of what happens over time in relation to particular attributes of plantings.

Changes in birds in plantings over time

By far the most work we have done on temporal patterns of change in wildlife has been on how populations of birds have changed in plantings in the past 20 years (Lindenmayer *et al.* 2016b). While much of this work is published in the scientific literature (and we provide citations to those articles where appropriate), other studies are ongoing and we only briefly cross-reference findings made to date for this unpublished research.

Several previous studies have found that the number of bird species in a planting increases over time. That is, older plantings tend to support more species (e.g. Greening Australia 2001; Martin *et al.* 2004; Lindenmayer *et al.* 2016b). The most likely reason for this is that older plantings often become more structurally complex and diverse as they age, being characterised by trees and some shrubs and regeneration at different heights with more leaf litter, sticks, and logs comprising ground cover (Munro *et al.* 2009). More complex vegetation can, in turn, support a greater range of niches for birds and thereby support more species. However, our long-term work in the South West Slopes of New South Wales has not shown such strong relationships between the age of plantings and the number of bird species (Lindenmayer *et al.* 2016b). Rather, bird species richness has remained relatively constant over time, although the composition of the bird assemblage has changed due to turnover in the different species that occur in plantings. That is, some bird species colonise plantings and live in them for some time before being replaced by new species. For example, migratory species are among those that are often absent from young plantings but colonise older plantings (Lindenmayer *et al.* 2016b). Other long-term studies have found there are changes in the composition of the bird community as plantings mature (Debus *et al.* 2017).

Analyses of the life history attributes of all birds show some interesting patterns. They show that both rare and common birds are declining across all vegetation types – that is, in plantings, natural regrowth and old growth woodland. Resident and migratory species are also declining in all vegetation types but markedly less so in plantings. Small birds are increasing in plantings but

Box 6.4. Changes in farming landscapes and changes in woodland birds over time

In general, there are more species of birds on farms and in landscapes with increased levels of native vegetation cover (Cunningham *et al.* 2014a; Cunningham *et al.* 2014b) (see Chapter 4). Some of our research has indicated that increasing the amount of native vegetation cover is the action that makes the single biggest contribution to promoting bird colonisation of plantings over time (Tulloch *et al.* 2016).

We have found that 12 species of birds showed an overall increase between 2002 and 2016. Of these 12 species, nine increased in plantings. These included several species of conservation concern such as the Rufous Whistler and White-browed Babbler (see Fig. 6.4).

Figure 6.3. Photo montage of some species which exhibited a significant **decline** in occurrence between 2000 and 2017, including in plantings. Clockwise from top left: Dusky Woodswallow, White-plumed Honeyeater and Grey Shrike-thrush. Dusky Woodswallow photo by Matthew Clancy, White-plumed Honeyeater and Grey Shrike-thrush by Graeme Chapman.

Figure 6.4. Photo montage of some species which exhibited a significant **increase** in occurrence between 2000 and 2017, including in plantings. Clockwise from left: White-winged Triller, Superb Fairy-wren and Rufous Whistler. Rufous Whistler and White-winged Triller by Graeme Chapman, Superb Fairy-wren by Beth Green.

not in other vegetation types, whereas large-bodied birds are declining in all vegetation types but especially in plantings. Further work has also shown that, relative to old growth woodlands, the bird assemblages in plantings tend to support more understorey foragers, omnivores, and cup- and dome-nesting species. Therefore, over time, plantings are helping to recruit different kinds of species that have different ecological roles relative to farms and agricultural landscapes where plantings are not being established. This has important implications for the key roles that birds can perform in farming landscapes, such as the control of insect pests.

Changes in weather and changes in birds

Australia has the world's most hyper-variable climate. Such variation in climate conditions can have marked effects on birds, including in plantings. Plantings are particularly important habitats for birds during periods of extreme temperature and reduced rainfall. In particular, small-bodied bird species and migratory species are significantly more likely to occur in plantings and similarly structured areas of natural regrowth than in old growth (non-restored) woodland. These same effects are not seen for large-bodied birds nor do they occur for non-migrant (i.e. resident) species. Interestingly, the value of plantings as habitat for small birds and migratory species is not apparent outside of drought periods and when temperatures are not extreme. Plantings (as well as patches of natural regrowth) are therefore used as a kind of 'climate refuge' during periods of extreme weather.

The densely spaced trees in plantings and regrowth provide cooler and more shaded micro-environments during periods of reduced rainfall and higher temperature. Such conditions may help birds lower their body temperature during extreme climate events (Gardner *et al.* 2016) and be particularly attractive for migratory species that have the added physiological challenges of long distance movements (Guillemete *et al.* 2016) before settling in a territory.

Changes in grazing regimes and changes in birds

As outlined in Chapter 5, fences are a critical form of infrastructure in restored vegetation because they help regulate the intensity, timing and frequency of grazing by domestic livestock. The effects of altered grazing regimes in plantings can be demonstrated when fences are removed. We have found that bird species richness declines markedly when formerly fenced plantings have the fences removed and grazing intensity increases (Lindenmayer *et al.* 2018). Changes in the litter layer on the woodland floor are important effects of grazing that, in turn, affect birds. These are key findings and they illustrate the value of grazing control, primarily through maintaining fences, including replacing them when they deteriorate (see Chapter 5).

Changes in reptiles over time

Relatively few species of reptiles respond positively to plantings, possibly because key habitat attributes like large logs and rocks are often missing from such places. They are often highly shaded and while that seems to be positive for some kinds of birds (like migratory species), this may create unsuitable micro-climatic conditions for reptiles. Despite these more general responses, some species such as the Boulenger's Skink and the Three-toed Earless Skink are found more often in plantings as they mature. Importantly, once these reptiles colonise plantings they can successfully breed there.

Changes in mammals over time

The long-term changes in occurrence that we have documented for birds and reptiles in plantings are not replicated for mammals like the Common Brushtail Possum and the Common Ringtail Possum. These and other species of arboreal marsupials such as the Squirrel Glider and Sugar Glider are very rarely found in plantings (Lindenmayer *et al.* 2017b). Moreover, over the 11-year period between 2002 and 2013, almost no plantings were colonised by these species. For example, plantings that were not occupied by Common Brushtail Possums or Common Ringtail Possums at the outset of our investigation in 2002 remained unoccupied throughout the subsequent decade. Both of these species (as well as all other species of possums and gliders in temperate woodlands) are largely dependent on cavities in large old trees for denning and nesting. It is likely that a lack of shelter

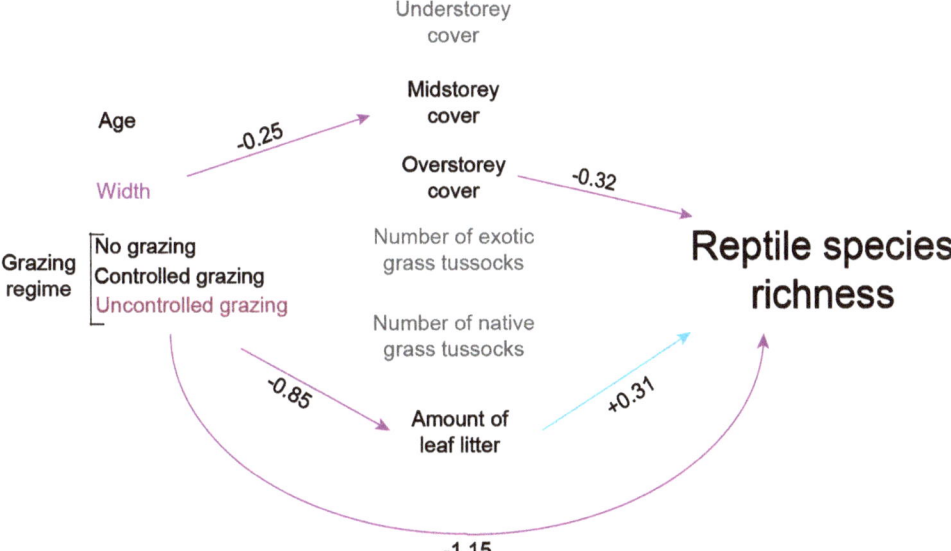

Figure 6.5. Links between grazing, changes in leaf litter and grazing-related effects on reptile species richness (Lindenmayer *et al.* 2018). The red arrows show negative relationships and the blue arrow positive relationships between variables.

Figure 6.6. Boulenger's Skink is a species more likely to colonise plantings as these areas mature. Photo by Damian Michael.

Box 6.5. Myth-busting: encouraging reptiles is bad

Some landholders have developed a negative impression of reptiles on their property, due to their reluctance to encourage the Eastern Brown Snake, which is a highly venomous reptile. However, there are many other reptiles that inhabit farm landscapes and present no danger to landholders. They range from an array of skinks and geckos to monitors (or goannas) and other kinds of (less venomous) snakes, as well as amphibians (frogs). Efforts to resist tidying up fallen timber on farms will benefit these other species and potentially reduce populations of the Eastern Brown Snake. Our anecdotal evidence suggests that the Eastern Brown Snake is less common on farms where fallen timber is left in plantings and woodland remnants. This could be because young Eastern Brown Snakes are preyed upon by other reptiles that benefit from fallen timber and other habitats on farms.

One myth is that with more fallen timber on a farm, many more sheep and cattle will die from snakebite. We suggest that snake bites of livestock extremely rare with the probability being so low that removing vast quantities of fallen timber is unwarranted.

sites in large old hollow-bearing trees is one of the key factors limiting the occupation of plantings by cavity-dependent arboreal marsupials. This suggests there will be a very long lag between when plantings are first established and when they become suitable for the Common Brushtail Possum and the Common Ringtail Possum as well as other species of possums and gliders. Our most recent data indicate a slow but nevertheless positive increase in the occupancy of plantings by arboreal marsupials, rising from less than 3% of sites in 2002 to ~10% of plantings in 2017.

The problems of a paucity of nesting sites might be mitigated when artificial cavities such as nest boxes are provided. Both the Common Brushtail Possum and the Common Ringtail Possum colonise plantings when nest boxes are established in them (Lindenmayer *et al.* 2016a). Other species of mammals using nest boxes in plantings are the Yellow-footed Antechinus and the Sugar Glider. Importantly, the use of nest boxes in plantings is significantly higher when such areas were well connected with tree cover to other plantings or patches of remnant eucalypt vegetation, most probably because it facilitates movements and subsequent colonisation (Lindenmayer *et al.* 2016a).

Key species of conservation concern like the Squirrel Glider can benefit from the establishment of well-designed nest boxes (Goldingay *et al.* 2015). For example, 60–80% of nest boxes established in a study located in the Lurg Hills, north-east Victoria, have shown signs of occupancy by Squirrel Gliders since 1996. Furthermore, during a single survey of nest boxes in 2009, this study detected Squirrel Gliders in 135 of 355 boxes, with a total of 257 animals recorded in various-sized groups (Thomas 2009). In addition, spotlighting and radio-tracking studies have found that these animals will forage in plantings, particularly older plantings. The Squirrel Glider also will use plantings as stepping stones to access paddock trees and patches of remnant woodland, especially when large old trees are flowering (as they become important sources of food) (Crane *et al.* 2014).

Plantings, time and long-term farm planning

As indicated by much of the commentary in this chapter, the value of plantings for wildlife changes over time. In the case of birds, a different assemblage of species is often found in young plantings relative to the assemblage characteristic of old plantings, recognising, of course, that other things like the composition, size, shape and context of plantings also influence wildlife occurrence. These temporal changes in animal occurrence mean that if an aim is to maximise the number of bird species on a farm, then a policy to add new areas of native vegetation over time might be appropriate. At the same time, it is important to keep some very old plantings as they become less suitable over time for 'problem' species like the Noisy Miner (Lindenmayer *et al.* 2016b) whose presence can have negative impacts on

Box 6.6. Be patient

Persistence and patience are not concepts that readily fit with much of modern Western culture. However, it can take some years for major gains in biodiversity to be realised from significant planting programs. They **do** occur and it is critical to be patient. Indeed, much of the biodiversity loss in Australia has occurred over 200 years of intensive and extensive land use, and reversing these declines will not happen in just a handful of years. Nevertheless, we have documented major changes in groups such as birds well within 10 years of a planting being established. There are many reasons to remain positive.

other bird species, particularly smaller-bodied birds (Grey *et al.* 1998) (Montague-Drake *et al.* 2011) (Mac Nally *et al.* 2012). Planning where plantings of different ages occur on a farm can be guided by a whole-of-farm plan, a key topic in the following chapter.

Summary

Few things in the ecological world stay the same. Many aspects of plantings, such as the structure of the vegetation and the identities of species of birds and reptiles, change over time. Some changes in wildlife over time are associated with changes in weather conditions such as the amount of rainfall and temperatures. Others are linked to changes in characteristics of the ground, shrub and overstorey layers of the vegetation. Some changes in biodiversity are positive, like the decline in the occurrence of the Noisy Miner as plantings age. However, even relatively old plantings lack some key resources such as large hollow-bearing trees (unless they are established around them in the first place) and very prolonged periods of time (sometimes exceeding a century) will be required for these features to develop. This underscores why plantings are not a valid substitute for old growth woodland on farms. The only way to document long-term changes in plantings is to monitor them carefully and on a repeated basis – a topic that we revisit in the following chapter.

7

Conclusions: creating a whole-of-farm plan and some thoughts on the future

Each of the preceding chapters in this book demonstrates that plantings are a critical part of the portfolio of vegetation assets on a farm. Plantings have many roles. Our focus in this book has been on biodiversity because that is what the group at The Australian National University has worked on intensively for the past two decades. But there are many other important roles of plantings on farms. When done well, plantings can be important for enhancing farm productivity, boosting lambing success, and providing shelter for stock during both high and low temperatures (Cleugh 2003).

As discussed in Chapter 2, different kinds of plantings can be important for different species of wildlife on farms. However, there are some general principles that are likely to make some plantings better environments for wildlife than others (Box 7.1). We recognise that it may not be possible to implement all of these things, but doing at least some of these activities will be far preferable to doing few or none.

The critical need for whole-of-farm plans

The previous chapter highlighted how plantings of different ages can provide habitat for different assemblages of birds, reptiles and other animals. One approach

Box 7.1. In a nutshell: what makes a good planting?

In general, the plantings that contribute best to the conservation of wildlife on farms are those that:

- are established around key features like large old paddock trees, logs and rocky areas
- are composed of a range of species of native plants, including shrubs and other kinds of vegetation
- are characterised by multiple layers of vegetation: ground cover, an understorey and an overstorey
- maintain mistletoe and other important features
- are established near other plantings or close to patches of remnant native vegetation
- are block-shaped and generally as wide as possible and/or physically connected to other plantings
- are well fenced to either exclude or limit intense grazing by domestic livestock, where possible avoiding the use of barbed wire on the top strand of fences
- are well managed so that weeds and pest animals such as rabbits are controlled.

to integrating agricultural production and biodiversity conservation is to continue to establish new areas of plantings over time, so that a farm has plantings of different ages distributed across a farmscape. Of course, this has limits as no farm can be all plantings and leave no reasonably sized areas for grazing and/or cropping. Some (but certainly not all) older plantings might be replaced with younger areas of more recently established plantings.

Determining how to timetable which plantings to establish where can be important and is best guided by the creation of a whole-of-farm plan. Prior to undertaking a whole-of-farm plan, it can be useful to conduct an audit of the environmental assets on a farm. What areas of a property support rocky outcrops, patches of remnant native woodland, and creeklines with some remnant trees? Where are the paddock trees and swards of native pasture? Where are the farm dams and the paddocks with the best lambing success? Are there road reserves or travelling stock reserves close to the farm boundaries? These and other assets can be mapped as part of a farm plan to help determine what the next options might be to promote some key aspects of farm restoration.

Another factor that may influence the development of a farm plan and hence key management decisions can be the presence of a particular species that a landowner wants to help conserve. Examples include the Bush Stone-curlew and Squirrel Glider, which some farmers have worked hard to plan for and then successfully conserve on their land.

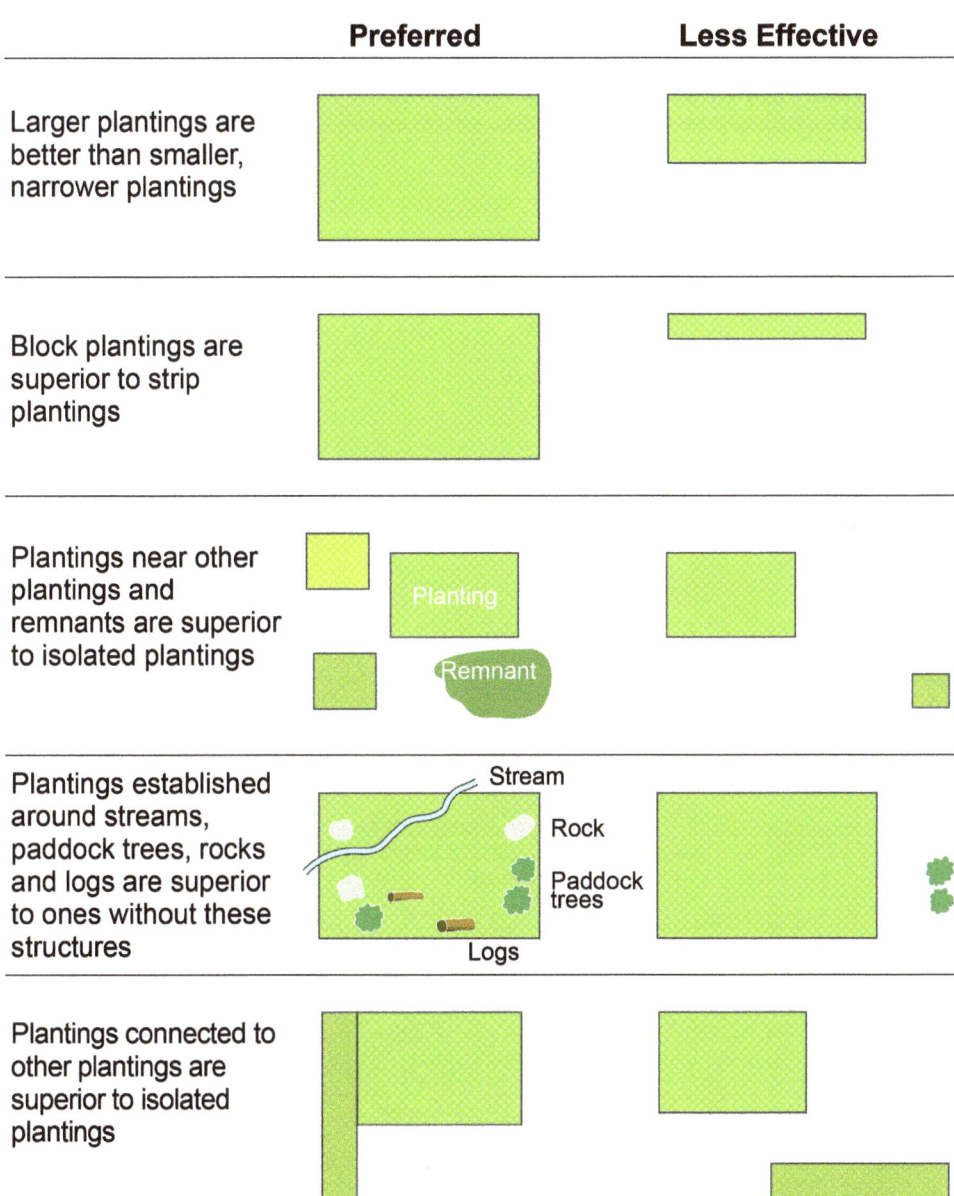

	Preferred	Less Effective
Larger plantings are better than smaller, narrower plantings		
Block plantings are superior to strip plantings		
Plantings near other plantings and remnants are superior to isolated plantings	Planting / Remnant	
Plantings established around streams, paddock trees, rocks and logs are superior to ones without these structures	Stream / Rock / Paddock trees / Logs	
Plantings connected to other plantings are superior to isolated plantings		

Figure 7.1. Diagrammatic representation of key features of plantings that are better for wildlife (redrawn from Lindenmayer *et al.* 2016c).

The summary points in Table 7.1 indicate that decisions about establishing plantings on a farm can be strongly influenced by what management has occurred previously on a farm and what vegetation assets are adjacent to a farm. For example, past decisions to leave particular areas of remnant vegetation on a farm can be influential because they are useful places (sometimes termed anchor points) near which to establish plantings. Similarly, watercourses can be good areas to target for replanting efforts because this is where revegetated areas tend to support the most species of native animals (Lindenmayer *et al.* 2010b). In addition, there is value in looking beyond a farm boundary; the biodiversity and other benefits of a given planting can be boosted if they are established near other plantings or areas of remnant vegetation on a neighbour's farm. This is because the planted area becomes

Table 7.1. Some key steps in creating a farm plan that includes plantings.

Action	Description
Create a map of the farm (and parts of the neighbouring farms) to show the location of natural assets.	The map should include patches of remnant vegetation, existing plantings, patches of native grassland rocky outcrops, paddock trees, streamlines, farm dams, paddocks and fences.
Map the natural assets adjacent to the farm.	The map should include travelling stock reserves, roadside reserves, remnant vegetation, and plants on the farms of the immediate neighbours.
Write down the objectives of management.	The objectives of a farm plan will require different kinds of management actions to be met. It is important to consider the different objectives carefully.
Consider the fencing options.	Fencing underlies many of the management interventions on a farm, particularly (1) establishing plantings and (2) modifying grazing pressure in existing remnants. The plan should highlight whether existing fences need replacing, where new fences are needed, and where old and/or new fences are best located relative to key assets like streamlines.
Consider the most appropriate best-management action.	The kind of management actions on a farm will often depend on what natural assets currently exist and what actions have already been taken as well as what funding and other opportunities are available.
Discuss what management practices might be coordinated with neighbours.	Some kinds of management are more effective if they are coordinated with landholders on adjacent properties. Fox baiting and weed control are two very good examples. However, the effectiveness of plantings on a farm often can be greater if they are established near other plantings and patches of remnant vegetation, including those on adjacent farms.
Seek advice on what support might be available to assist with replanting or other kinds of management.	Often there will be resources available to assist landowners establish and manage restored native vegetation.
Record what has been done where, and when, and how.	It is important to document what management was done where and how it was done. Photos of what areas used to look like relative to now are also important to demonstrate the changes that have occurred.

functionally larger for some species. Indeed, extensive analysis of two decades of long-term data has indicated that increasing the overall amount of native vegetation cover in a landscape is the single most effective action for increasing the occupancy of woodland patches, including plantings (Tulloch *et al.* 2016).

Creating farm plans has many other benefits because, when done well, such a plan can significantly improve productivity and ultimately profitability. It is therefore a valuable opportunity to plan what the farm will look like in the future and, over time, judge progress against what was planned. Maps and aerial photographs are particularly useful in guiding farm plans and are especially satisfying to peruse several years after active planting programs have occurred. All of the farms on which we have long-term monitoring sites, and which have been successfully (and profitably) managed, have well-developed farm plans.

Box 7.2. 'Balancing the portfolio' of vegetation assets

In several parts of this book we have discussed how plantings, stands of regenerating woodland, and patches of old growth remnant woodland are different elements of a portfolio of vegetation assets on a farm. We have also discussed how there is no such thing as a perfect planting (see Box 2.4 in Chapter 2). It is therefore important to make sure that, over time, there is a 'balanced portfolio' of vegetation assets on a farm. In the case of plantings, this may mean ensuring that some plantings are wide and block-shaped, and others are sparsely stocked with trees in contrast to those that have a dense spacing of trees and shrubs. Taking stock of what already occurs on a farm will help guide any decisions about what the next appropriate management actions might be (Ikin *et al.* 2016b).

Box 7.3. Scales of effect and 'connections' in the landscape

Much of this book has focused on the importance of plantings at the site level for farm wildlife. However, our work has shown that plantings can have far broader effects beyond individual sites where restoration is done, and have positive effects at larger paddock, farm and even landscape scales. This is one of the reasons why whole-of-farm planning can be so important. For example, plantings on farms can interact with other kinds of native vegetation to boost overall farm-level bird biodiversity (Cunningham *et al.* 2008). Similarly, overall bird species richness in agricultural landscapes is elevated with increasing levels of native vegetation cover (Cunningham *et al.* 2014a; Cunningham *et al.* 2014b), which can occur through coordinated restoration programs that aim to increase the number of planted sites (and hence the area of plantings) as well as allow for natural regeneration.

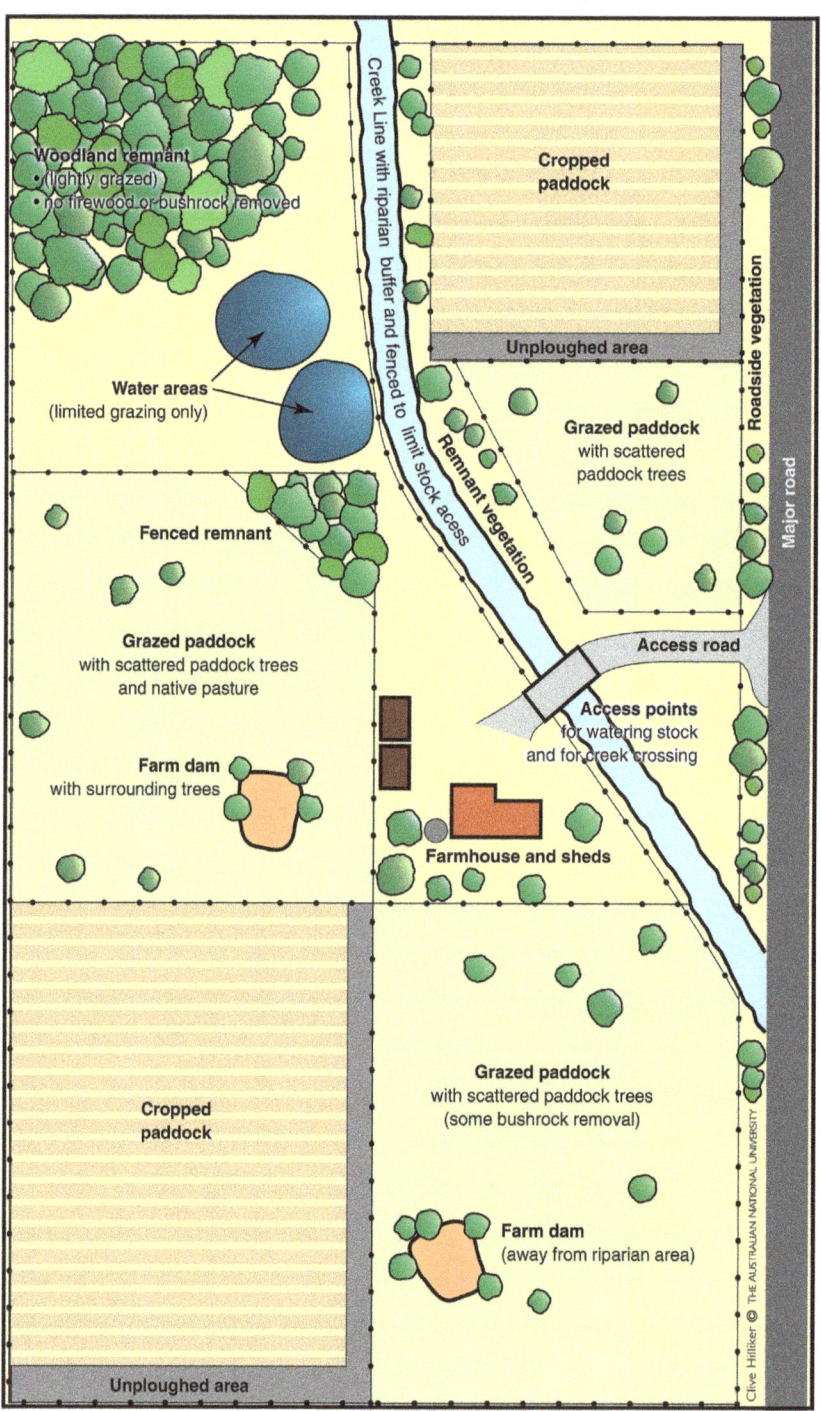

Figure 7.2. A crude map of a farm showing the location of key natural assets that form the foundation for restoration programs and enhancing farm management practices.

Good monitoring is essential

The only way to truly determine how things are changing in plantings is to monitor them. Without monitoring there is no way of being able to tell whether establishing plantings or other management interventions like weed removal or grazing control have been effective. Indeed, the only reason we have been able to write this book with confidence is that we have conducted such major long-term monitoring programs in restored areas and regrowth and old growth woodlands.

Best practice monitoring requires careful planning, good experimental design and dedicated survey efforts that are repeated consistently over many years. Countless books and scientific articles have been written about this topic, and we will not repeat even a tiny fraction of that material here. Nonetheless, it is always important to seek some expert help in the best ways to establish a monitoring program to ensure that the data being gathered are useful to determine if things are changing, how they are changing and, if possible, why they are changing. There are some basic principles that are essential to an effective monitoring program:

1. Determine why you want to monitor. For example, is it to see if the trees survive in your planted areas or to determine how many bird species have colonised a restored area?
2. Work out what you want to measure in the monitoring program (e.g. the number of dead trees, the range of species of birds, or the abundance of lizards).
3. Design the monitoring program so that you can compare the trees or birds (or other entities that you are measuring) between places that have been subject to different kinds of management. For example, planted areas may be compared with paddocks and remnant old growth woodland, or large block plantings might be contrasted with long narrow plantings. The design of a monitoring program can be particularly strong if it is started before a management intervention is started (e.g. a year or so before plantings are established). This allows the level of change to be quantified before and then after management.
4. Make sure there are multiple sites in each category of management (e.g. a number large plantings to then compare to several small plantings). This helps account for variation caused by something special that has occurred at one particular site (like it being the former location of a sheep dip).
5. Work out the kind of survey unit that will be same for all sites (e.g. a transect of a set length or a plot of a fixed size).
6. Work out the best time to survey sites. For example, repeated winter surveys for butterflies in plantings will not be particularly productive; spring is the appropriate time to monitor this group.
7. Continue to collect data for several years (the longer the better, given how variable Australian weather and ecosystems can be). Be patient and don't give

up if you don't see immediate results as changes in planted areas take time (see Box 6.6).

8. Make sure that the data from the things being measured are carefully stored so that they can used for analysis of changes over time.

If these steps seem overwhelming, it is because good and effective monitoring can often be a specialised task. Partnerships between landholders and people with expertise in monitoring (such as scientists from universities) can help ensure that monitoring is effective. In addition, a lot of monitoring is being done under citizen science programs, and it can be valuable if it is guided by a robust scientific framework.

Even if it proves impossible to conduct the kinds of monitoring with the ingredients outlined above, there is nevertheless enormous value in doing some simple things such as writing down what management actions have been done, where and when and using what methods (e.g. using tubestock versus direct seeding to establish a planting). Taking photos is also valuable for showing the amount of change that has occurred on parts of a farm with a successful restoration program.

Partnerships are necessary

No-one has all the answers. Almost everyone has tried something that has worked and other things that have failed. Sharing experiences of successes and failures is critical to help others avoid making the same mistakes. Part of developing better partnerships is to foster greater levels of communication between interested parties, between resource managers and farmers, between scientists and farmers and resource managers, and between farmers, scientists and policy makers. Communication is one of the key reasons we wrote this book – many of the people we work with ask us repeatedly about new insights and what the science is telling us.

Plantings and farming in a rapidly changing climate

Very few rational, clear-thinking scientists deny the overwhelming evidence for rapid climate change. Those that claim there is no evidence of climate change have failed to understand that thousands of carefully designed and expertly executed scientific studies are published on this topic every year – using the same general scientific methodology as employed in medical research, engineering studies and all manner of rigorous scientific endeavour.

Some ill-informed commentators claim that climate models do not work, and have chosen instead to follow a head-in-the-sand approach and assert climate change is not happening. It is true that climate models have their limitations – as do all models, including those for more simple entities like crop prediction. Yet the empirical data clearly show that levels of carbon dioxide in the atmosphere are increasing rapidly, and that there are major changes in populations of plants and animals. Animals are changing in body size, some are breeding earlier, and others are colonising new environments. An enormous number of other environmental and other changes, such as magnified droughts, more frequent and more intense wildfires, and increased rates of tree mortality, are occurring. What does this mean for vegetation restoration and, in particular, establishing plantings on farms? The answers to this question are far from straightforward. However, there are some important tactics and strategies that can be employed in vegetation restoration in a more volatile climate future.

First, native vegetation cover is critical, particularly as several studies now show that drought effects tend to be magnified in regions where extensive tree clearing has taken place (e.g. McAlpine et al. 2007; Steffen et al. 2009). Therefore, it is important to not only maintain existing vegetation cover but to also employ planting programs to increase the amount of vegetation that characterises farms and broader agricultural landscapes.

Second, use tree species that are regionally prevalent to establish plantings. This is a useful rule-of-thumb because many of the dominant tree species in native woodlands are already widespread species that persist over relatively large areas and should, at least in theory, be able to survive in environments subject to moderate amounts of short-medium changes in climate.

Third, the kinds of trees and the environmental conditions into which they are being established should be carefully considered. This is crucial for two reasons: (1) the germination and early growth stages of tree development are often the most vulnerable parts of a tree's life cycle, and (2) the conditions under which existing mature trees first established (sometimes many centuries ago) may have been very different from those which exist now and therefore may, in turn, be very challenging for growth. In these cases, steps need to be taken (like limiting grazing pressure and implementing appropriate site preparation) to ensure that trees have the greatest possible chance of becoming established.

Fourth, the reasons for establishing plantings are even more compelling under conditions of increased rates of climate change. They are not only critically important because of the shade they provide and hence for the maintenance of the health of livestock (particularly in terms of reducing water demand and increasing weight gain), but our work shows that they also act as key refuges for bird biodiversity during the depressed levels of rainfall and elevated temperatures that characterise droughts. Such conditions will become more frequent and extreme as a result of a changing climate (Steffen et al. 2009).

More work to do

In this book, we have reported the results of some of our studies completed over the past five years. These studies, in turn, build on the previous 15 years of research and monitoring. Together these insights have been melded into a body of work which, sadly, has few parallels elsewhere in the world, including in Australia. But there is still more work to be done. This is not a lame call for more research simply for the sake of it and keeping scientists employed. Rather, there is still an enormous amount of new knowledge needed to help improve the integration of wildlife conservation into agricultural production. Such improvements are critical given that, for example, livestock grazing is the biggest use of land not only on the Australian continent but worldwide (FAO 2009; Williams and Price 2010). We therefore must find ways to ensure that conservation values can be maintained where agricultural production takes place to supply the food and fibre needed to support a rapidly increasing human population.

As an example, we presently have few insights into how to best use plantings to maintain connectivity among dispersed populations of species of plants and animals. That is, how can we best design where plantings are located to facilitate the movement of species across agricultural landscapes? In many cases, species can move without physically connecting planted areas with patches of remnant native vegetation, but new plantings nevertheless do need to be established in particular places so that they act as climate refuges for native wildlife and some native plant species. A key challenge in the coming three to five years will be developing mapping tools that indicate where the best places are to maximise the biodiversity gains that can be generated from restoration programs. These maps need to be based on information on the habitat requirements of plants and animals, but also on knowledge of how far species can move across farming landscapes to colonise new habitats and establish new populations or prevent existing ones from declining or going locally extinct.

A further important research challenge is to identify ways to determine how to optimise the size, shape and location of plantings so they are effective for biodiversity, act as high-quality shelterbelts, and also store large amounts of carbon. That is, there is much work to be done to maximise the various co-benefits from active replanting programs on farms. This is important because if a carbon accounting methodology can be developed for Australian agricultural landscapes, then there is enormous potential for farmers to generate additional forms of income from the role they will play in tackling climate change, and at the same time contributing to farm productivity and profitability, as well as addressing issues with farm wildlife conservation.

Another area where new work is needed is how to accelerate the recovery of populations of large old trees, which are keystone structures in farming landscapes.

Figure 7.3. A large old Red Box tree. Climatic conditions when this 300+-year-old tree may well have been markedly different from those that currently exist. Photo by Mason Crane.

Figure 7.4. Replanted areas. Plantings are critical micro-refugia for native species such as birds, particularly during extreme droughts and their associated high temperatures and periods of low rainfall. Photo by Tabitha Boyer.

This is an urgent task given how many large old trees have been cleared in the past and also how long it can take for new trees to be recruited when existing trees are lost. Allied with this work is the need to develop better designs for nest boxes to substitute for the hollow-nesting species that are otherwise unable to persist without access to large old trees. New kinds of nest boxes are needed that both exclude pest species and last for much longer than existing types of boxes (that are often unsuitable for occupancy within 10 years). Hollow augmentation through the creation of artificial hollows with chainsaws requires more monitoring and research to assess the effectiveness of this approach.

In many cases, the most important insights are generated from very long-term studies, particularly in Australia. This is not only because of the continent's hyper-variable climate but also the fact that some changes (like the suitability of plantings as habitat for birds and reptiles) can take a long time to manifest. We are adamant that funds for, and ways to, continue the legacy of the past 20 years of research and monitoring must be found, especially as our extensive network of field sites is a critical part of the Australia's environmental infrastructure. Without it we would have no way of telling if things are improving or if the management interventions that farmers have implemented on their farms (including establishing

Box 7.4. When politics breaks the critical links between good management and good science

'We already know how to stick trees in the ground, scientists just need to get out of the way and let practical people get on with the business of establishing tree plantings.'

The words at the start of this box paraphrase the comments by the senior environmental advisors to Senator Robert Hill (then the Commonwealth Environment Minister) in late 1996 just after the election of the Howard Government, and when the Natural Heritage Trust first commenced.

Actually, at the time of this quote, remarkably little was known about what makes an effective tree planting. Indeed, this is a classic example of Brandolini's Bullshit Asymmetry Principle, 'The amount of energy necessary to refute bullshit is an order of magnitude bigger than to produce it'. This book outlines some of the important new scientific insights generated from the last 20 years of intensive scientific research on how to establish ecological effective restored or replanted areas. Looking back, we speculate that hundreds of millions of dollars of precious taxpayer funds were wasted because restoration programs were either not underpinned by good science and/or there was no robust monitoring to help determine if actions to restore native vegetation cover were actually effective (or not). We now know better that science and management must go hand in hand, and that farmers and other kinds of land managers can work with researchers to maximise the benefits of revegetation programs and the sustainable management of agricultural land.

plantings and adopting various forms of grazing control) have been successful. Despite the importance of long-term research and monitoring, they lack support from the Australian Government. For example, the nation's Long-term Ecological Research Network is no longer being funded, even though the network encompasses some of the most important long-term programs nationwide. In addition, the monitoring program for the Environmental Stewardship Program has been terminated in a way that will make it impossible to tell if ongoing management is effective. These are highly retrogressive steps. Nevertheless, other long-term research and monitoring will undoubtedly produce important new insights, some of which we cannot even presently imagine. Our hope is that perhaps in another five to seven years' time, we can produce an updated version of this book (or one similar to it) that documents a new set of exciting and important discoveries around how we can improve the value of plantings for farm wildlife and, in doing so, help better integrate conservation values with agricultural production.

Figure 7.5. She-oak woodland with Acacia understorey, Serpentine Hill, Coolac area, New South Wales. Photo by Esther Beaton.

Concluding comments

'When I was a kid in the 1940s, if I had been naughty then dad told to me get outside and chop down a tree. Now it is the opposite – planting trees is important and there's far more vegetation than there used to be ...'

We finish this book where we started, with the quote from Ross Cunningham. That is, many farms and landscapes (and also many farmers) in agricultural south-eastern Australia have changed in significant ways over the past few decades. Many farms now support far more trees than they used to. This is an inherently good thing. Through good monitoring and other kinds of science, we can document the positive effects these changes have had, not only on biodiversity but also on farm profitability and farmer mental health. This does not mean that our work is done. Far from it. Learning should be a life-long enterprise and continued across generations. There is much to learn about how to restore farms and how to best integrate agricultural production and on-farm conservation practices. Major programs like the Natural Heritage Trust and Caring for our Country wasted hundreds of millions of dollars on management that was done in the absence of science and monitoring to guide effective practice (see Box 7.4). It is our sincere hope that the need for partnerships between landholders, other resource managers and scientists will persist for decades and generations to come to ensure that agricultural production and on-farm conservation can be integrated.

References

Barrett GW, Ford HA, Recher HF (1994) Conservation of woodland birds in a fragmented rural landscape. *Pacific Conservation Biology* **1**, 245–256. doi:10.1071/PC940245

Barrett GW, Freudenberger D, Drew A, Stol J, Nicholls AO, Cawsey EM (2008) Colonisation of native tree and shrub plantings by woodland birds in an agricultural landscape. *Wildlife Research* **35**, 19–32. doi:10.1071/WR07100

Barton PS, Manning AD, Gibb H, Wood JT, Lindenmayer DB, Cunningham SA (2011) Experimental reduction of native vertebrate grazing and addition of logs benefit beetle diversity at multiple scales. *Journal of Applied Ecology* **48**, 943–951. doi:10.1111/j.1365-2664.2011.01994.x

Barton PS, Cunningham SA, Lindenmayer DB, Manning AD (2013) The role of carrion in maintaining biodiversity and ecological processes in terrestrial ecosystems. *Oecologia* **171**, 761–772. doi:10.1007/s00442-012-2460-3

Barton PS, Sato CF, Kay GM, Florance D, Lindenmayer DB (2016) Effects of environmental variation and livestock grazing on ant community structure in temperate eucalypt woodland. *Insect Conservation and Diversity* **9**, 124–134. doi:10.1111/icad.12151

Batáry P, Dicks LV, Kleijn D, Sutherland WJ (2015) The role of agri-environment schemes in conservation and environmental management. *Conservation Biology* **29**, 1006–1016. doi:10.1111/cobi.12536

Bell I, Priestley T (1998) Management of stock access to the riparian zone. Department of Primary Industries, Water and Environment, Hobart, Tasmania.

Bird PR, Jackson TT, Kearney GA, Williams KW (2002) Effect of two tree windbreaks on adjacent pastures in south-western Victoria, Australia. *Australian Journal of Experimental Agriculture* **42**, 809–830. doi:10.1071/EA02016

Bond S (2012) *Avian Responses to Farm Plantings*. The Australian National University, Canberra.

Burns EL, Zammit C, Attwood SJ, Lindenmayer DB (2016) The Environmental Stewardship Program: Lessons on creating long-term agri-environment schemes. In *Learning from Agri-environment Schemes in Australia: Investing in Biodiversity and Other Ecosystem Services on Farms*. (Eds D Ansell, F Gibson, D Salt) pp. 33–51. ANU E-Press, Canberra.

Cleugh H (2003) *Trees for Shelter: A Guide to Using Windbreaks on Australian Farms*. Rural Industries Research and Development Corporation, Canberra, Australia.

Cooper CB, Walters JR, Ford H (2002) Effects of remnant size and connectivity on the response of

Brown Treecreepers to habitat fragmentation. *Emu* **102**, 249–256. doi:10.1071/MU01007

Crane M, Cunningham RB, Lindenmayer DB (2014) The value of countryside elements in the conservation of a threatened arboreal marsupial *Petaurus norfolcensis* in agricultural landscapes of south-eastern Australia – the disproportional value of scattered trees. *PLoS One* **9**, e107178. doi:10.1371/journal.pone.0107178

Crane M, Lindenmayer DB, Cunningham RB, Stein J (2016) The effect of wildfire on scattered trees, 'keystone structures', in agricultural landscapes. *Austral Ecology.* doi:10.1111/aec.12414.

Crane M, Montague-Drake RM, Cunningham RB, Lindenmayer DB (2008) The characteristics of den trees used by the Squirrel Glider (*Petaurus norfolcensis*) in temperate Australian woodlands. *Wildlife Research* **35**, 663–675.

Cunningham RB, Lindenmayer DB, Crane M, Michael D, MacGregor C (2007) Reptile and arboreal marsupial response to replanted vegetation in agricultural landscapes. *Ecological Applications* **17**, 609–619. doi:10.1890/05-1892

Cunningham RB, Lindenmayer DB, McGregor C, Crane M, Michael D (2008) The combined effects of remnant vegetation and replanted vegetation on farmland birds. *Conservation Biology* **22**, 742–752. doi:10.1111/j.1523-1739.2008.00924.x

Cunningham RB, Lindenmayer DB, Barton P, Ikin K, Crane M, Michael D, Okada S, Gibbons P, Stein J (2014a) Cross-sectional and temporal relationships between bird occupancy and vegetation cover at multiple spatial scales. *Ecological Applications* **24**, 1275–1288. doi:10.1890/13-0872.1

Cunningham RB, Lindenmayer DB, Crane M, Michael D, Barton PS, Gibbons P, Okada S, Ikin K, Stein JAR (2014b) The law of diminishing returns: woodland birds respond to native vegetation cover at multiple spatial scales and over time. *Diversity & Distributions* **20**, 59–71. doi:10.1111/ddi.12145

Debus S, Martin WK, Lemon JM (2017) Changes in woodland bird communities as replanted woodland matures. *Pacific Conservation Biology* **23**(4), 359–371.

Dorrough J, Moxham C (2005) Eucalypt establishment in agricultural landscapes and implications for landscape-scale restoration. *Biological Conservation* **123**, 55–66.

Driscoll DA, Lindenmayer DB (2009) Empirical test of metacommunity theory using an isolation gradient. *Ecological Monographs* **79**, 485–501. doi:10.1890/08-1114.1

Duncan DH, Dorrough J (2009) Historical and current land use shape landscape restoration options in the Australian wheat and sheep farming zone. *Landscape and Urban Planning* **91**, 124–132. doi:10.1016/j.landurbplan.2008.12.007

FAO (2009) *The State of Food and Agriculture 2009. Livestock in the Balance.* FAO, Rome.

Fischer J, Lindenmayer DB (2002) The conservation value of paddock trees for birds in a variegated landscape in southern New South Wales. 2. Paddock trees as stepping stones. *Biodiversity and Conservation* **11**, 833–849. doi:10.1023/A:1015318328007

Fischer J, Lindenmayer DB, Montague-Drake R (2008) The role of landscape texture in conservation biogeography: a case study on birds in south-eastern Australia. *Diversity & Distributions* **14**, 38–46. doi:10.1111/j.1472-4642. 2007.00411.x

Fischer J, Stott J, Zerger A, Warren G, Sherren K, Forrester RI (2009) Reversing a tree regeneration crisis in an endangered ecoregion. *Proceedings of the National Academy of Sciences of the United States of America* **106**, 10386–10391. doi:10.1073/pnas.0900110106

Foster CN, Barton PS, Wood JT, Lindenmayer DB (2015) Interactive effects of fire and large herbivores on web-building spiders. *Oecologia* **179**, 237–248. doi:10.1007/s00442-015-3323-5

Foster CN, Barton PS, Sato CF, Wood JT, MacGregor CI, Lindenmayer DB (2016) Herbivory and fire interact to affect forest understory habitat, but not its use by small vertebrates. *Animal Conservation* **19**, 15–25. doi:10.1111/acv.12210

Gardner JL, Symonds MR, Joseph L, Ikin K, Stein J, Kruuk LE (2016) Spatial variation in avian bill size is associated with humidity in summer among Australian passerines. *Climate Change Responses* **3**, 1–11. doi:10.1186/s40665-016-0026-z

Gibb H, Cunningham SA (2010) Revegetation of farmland restores function and composition of epigaeic beetle assemblages. *Biological Conservation* **143**, 677–687. doi:10.1016/j.biocon.2009.12.005

Gibbons P, Lindenmayer DB (2002) *Tree Hollows and Wildlife Conservation in Australia*. CSIRO Publishing, Melbourne.

Gibbons P, van Bommel L, Gill AM, Cary GJ, Driscoll DA, Ross A, Bradstock RA, Knight E, Moritz MA, Stephens SL, Lindenmayer DB (2012) Land management practices associated with house loss in wildfires. *PLoS One* **7**, e29212. doi:10.1371/journal.pone.0029212

Goldingay RL, Rueegger NN, Grimson MJ, Taylor BD (2015) Specific nest box designs can improve habitat restoration for cavity-dependent arboreal mammals. *Restoration Ecology* **23**, 482–490. doi:10.1111/rec.12208

Gould S, Mackey B (2015) Site vegetation characteristics are more important than landscape context in determining bird assemblages in revegetation. *Restoration Ecology* **23**, 670–680. doi:10.1111/rec.12222

Greening Australia (2001) *Bringing Birds Back: A Glovebox Guide for Bird Identification and Habitat Restoration in ACT and SE NSW*. Greening Australia, ACT and SE NSW, Canberra.

Greening Australia (2003) *Revegetation Techniques*. Greening Australia Victoria, Melbourne.

Grey MJ, Clarke MF, Loyn RH (1998) Influence of the Noisy Miner *Manorina melanocephala* on avian diversity and abundance in remnant

Grey Box woodland. *Pacific Conservation Biology* **4**, 55–69. doi:10.1071/PC980055

Guillemete M, Woakes AJ, Larochelle J, Polymeropoulos ET, Granbois J-M, Butler PJ, Pelletier D, Frapell PB, Portugal SJ (2016) Does hyperthermia constrain flight duration in a short-distance migrant? *Philosophical Transactions of the Royal Society of London. Series B, Biological Sciences* **371**, 20150386. doi:10.1098/rstb.2015.0386

Hazell D, Cunnningham R, Lindenmayer D, Mackey B, Osborne W (2001) Use of farm dams as frog habitat in an Australian agricultural landscape: factors affecting species richness and distribution. *Biological Conservation* **102**, 155–169. doi:10.1016/S0006-3207(01)00096-9

Hazell D, Hero JM, Lindenmayer DB, Cunningham RB (2004) A comparison of constructed and natural habitat for frog conservation in an Australian agricultural landscape. *Biological Conservation* **119**, 61–71. doi:10.1016/j.biocon.2003.10.022

Heard GW, Black D, Robertson P (2004) Habitat use by the inland carpet python (*Morelia spilota metcalfei*: Pythonidae): seasonal relationships with habitat structure and prey distribution in a rural landscape. *Austral Ecology* **29**, 446–460. doi:10.1111/j.1442-9993.2004.01383.x

Holland GJ, Clarke MF, Bennett AF (2017) Prescribed burning consumes key forest structural components: implications for landscape heterogeneity. *Ecological Applications* **27**, 845–858. doi:10.1002/eap.1488

Howland B, Stojanovic D, Gordon IJ, Manning AD, Fletcher D, Lindenmayer DB (2014) Eaten out of house and home: impacts of grazing on ground-dwelling reptiles in Australian grasslands and grassy woodlands. *PLoS One* **9**, e105966. doi:10.1371/journal.pone.0105966

Ikin K, Yong DL, Lindenmayer DB (2016a) Effectiveness of woodland birds as taxonomic surrogates in conservation planning for biodiversity on farms. *Biological Conservation* **204**, 411–416. doi:10.1016/j.biocon.2016.11.010

Ikin K, Tulloch A, Gibbons P, Ansell D, Seddon J, Lindenmayer DB (2016b) Evaluating complementary networks of restoration plantings for landscape-scale occurrence of temporally dynamic species. *Conservation Biology* **30**, 1027–1037. doi:10.1111/cobi.12730

Jenkins M, Collins L, Penman T, Zylstra P, Horsey B, Bradstock R (2016) Environmental values and fire hazard of eucalypt plantings. *Ecosphere* **7**, e01528. doi:10.1002/ecs2.1528

Kavanagh RP, Stanton MA, Herring MW (2007) Eucalypt plantings on farms benefit woodland birds in south-eastern Australia. *Austral Ecology* **32**, 635–650. doi:10.1111/j.1442-9993.2007.01746.x

Kay GM, Mortelliti A, Tulloch A, Barton P, Florance D, Cunningham SA, Lindenmayer DB (2017) Effects of past and present livestock grazing on herpetofauna in a landscape-scale experiment. *Conservation Biology* **31**, 446–458. doi:10.1111/cobi.12779

Kinross C (2004) Avian use of farm habitats, including windbreaks, on the New South Wales tablelands. *Pacific Conservation Biology* **10**, 180–192. doi:10.1071/PC040180

Kinross C, Nicol H (2008) Responses of birds to the characteristics of farm windbreaks in central New South Wales, Australia. *Emu* **108**, 139–152. doi:10.1071/MU06024

Lindenmayer DB (2017) Conserving large old trees as small natural features. *Biological Conservation* **211**, 51–59. doi:10.1016/j.biocon.2016.11.012

Lindenmayer DB, Laurance W (2016) The ecology, distribution, conservation and management of large old trees. *Biological Reviews of the Cambridge Philosophical Society.* doi:10.1111/brv.12290.

Lindenmayer DB, Crane M, Michael D, MacGregor C, Cunningham RB (2005) *Woodlands: A Disappearing Landscape.* CSIRO Publishing, Melbourne.

Lindenmayer DB, Cunningham R, Crane M, Michael D, Montague-Drake R (2007) Farmland bird responses to intersecting replanted areas. *Landscape Ecology* **22**, 1555–1562. doi:10.1007/s10980-007-9156-9

Lindenmayer DB, Bennett AF, Hobbs RJ (Eds) (2010a) *Temperate Woodland Conservation and Management.* CSIRO Publishing, Melbourne.

Lindenmayer DB, Knight EJ, Crane MJ, Montague-Drake R, Michael DR, MacGregor CI (2010b) What makes an effective restoration planting for woodland birds? *Biological Conservation* **143**, 289–301. doi:10.1016/j.biocon.2009.10.010

Lindenmayer DB, Archer S, Barton P, Bond S, Crane M, Gibbons P, Kay G, MacGregor C, Manning A, Michael D, Montague-Drake R, Munro N, Muntz R, Okada S, Stagoll K (2011) *What Makes a Good Farm for Wildlife?* CSIRO Publishing, Melbourne.

Lindenmayer DB, Northrop-Mackie AR, Montague-Drake R, Crane M, Michael D, Okada S, Gibbons P (2012a) Not all kinds of revegetation are created equal: revegetation type influences bird assemblages in threatened Australian woodland ecosystems. *PLoS One* **7**, e34527. doi:10.1371/journal.pone.0034527

Lindenmayer DB, Wood J, Montague-Drake R, Michael D, Crane M, Okada S, MacGregor C, Gibbons P (2012b) Is biodiversity management effective? Cross-sectional relationships between management, bird response and vegetation attributes in an Australian agri-environment scheme. *Biological Conservation* **152**, 62–73. doi:10.1016/j.biocon.2012.02.026

Lindenmayer DB, Zammit C, Attwood SA, Burns E, Shepherd CL, Kay GE, Wood J (2012c) A novel and cost-effective monitoring approach for outcomes in an Australian biodiversity conservation incentive program. *PLoS One* **7**, e50872. doi:10.1371/journal.pone.0050872

Lindenmayer DB, Lane PW, Westgate MJ, Crane M, Michael D, Okada S, Barton PS (2014) An empirical assessment of the focal species hypothesis. *Conservation Biology* **28**, 1594–1603. doi:10.1111/cobi.12330

Lindenmayer DB, Crane M, Blanchard W, Okada S, Montague-Drake R (2016a) Do nest boxes in restored woodlands promote the conservation of hollow-dependent fauna? *Restoration Ecology* **24**, 244–251. doi:10.1111/rec.12306

Lindenmayer DB, Lane PW, Barton PS, Crane M, Ikin K, Michael DR, Okada S (2016b) Long-term bird colonization and turnover in restored woodlands. *Biodiversity and Conservation* **25**, 1587–1603. doi:10.1007/s10531-016-1140-8

Lindenmayer DB, Michael D, Crane M, Okada S, Florance D, Barton PKI (2016c) *Wildlife Conservation in Farm Landscapes.* CSIRO Publishing, Melbourne.

Lindenmayer DB, Crane M, Evans MC, Maron M, Gibbons P, Bekessy S, Blanchard W (2017a) The anatomy of a failed offset. *Biological Conservation* **210**, 286–292. doi:10.1016/j.biocon.2017.04.022

Lindenmayer DB, Mortelliti A, Ikin K, Pierson J, Crane M, Michael D, Okada S (2017b) The vacant planting: limited influence of habitat restoration on patch colonization patterns by arboreal marsupials in south-eastern Australia. *Animal Conservation* **20**, 294–304. doi:10.1111/acv.12316

Lindenmayer DB, Blanchard W, Crane M, Michael D, Sato S (2018) Are biodiversity benefits of vegetation restoration undermined by livestock grazing? *Restoration Ecology* (Online early). doi:10.1111/rec.12676

Lindenmayer DB, Blanchard W, Crane M, Michael D, Florance D (in press) Size or quality. What matters in vegetation restoration for bird biodiversity in endangered temperate woodlands? *Austral Ecology.*

Louv R (2005) *Last Child in the Woods: Saving Our Children from Nature-Deficit Disorder.* Algonquin Books of Chapel Hill, Chapel Hill, North Carolina.

Mac Nally R, Bowen M, Howes A, McAlpine CA, Maron M (2012) Despotic, high-impact species and the subcontinental scale control of avian assemblage structure. *Ecology* **93**, 668–678. doi:10.1890/10-2340.1

Mackey BG, Prentice IC, Steffen W, House JI, Lindenmayer DB, Keith H, Berry S (2013) Untangling the confusion around land carbon science and climate change mitigation policy. *Nature Climate Change* **3**, 552–557. doi:10.1038/nclimate1804

Majer JD, Recher HF, Graham R, Watson A (2001) *The Potential of Revegetation Programs to Encourage Invertebrates and Insectivorous Birds.* Curtin University, Perth.

Manning AD, Fischer J, Lindenmayer DB (2006) Scattered trees are keystone structures – implications for conservation. *Biological Conservation* **132**, 311–321. doi:10.1016/j.biocon.2006.04.023

Manning AD, Lindenmayer DB, Cunningham RB (2007) A study of coarse woody debris volumes in two box-gum grassy woodland reserves in the Australian Capital Territory. *Ecological Management & Restoration* **8**, 221–224. doi:10.1111/j.1442-8903.2007.00371.x

Manning AD, Gibbons P, Fischer J, Oliver D, Lindenmayer DB (2013) Hollow futures? Tree decline, lag effects and hollow-dependent species. *Animal Conservation* **16**, 395–403. doi:10.1111/acv.12006

Martin TG, McIntye S (2007) Impacts of livestock grazing and tree clearing on birds of woodland and riparian habitats. *Conservation Biology* **21**,

504–514. doi:10.1111/j.1523-1739. 2006.00624.x

Martin WK, Eyears-Chaddock M, Wilson BR, Lemon J (2004) The value of habitat reconstruction to birds at Gunnedah, New South Wales. *Emu* **104**, 177–189. doi:10.1071/MU02053

Massey C (2017) *Call of the Reed Warbler: A New Agriculture – A New Earth*. University of Queensland Press, Brisbane.

McAlpine CA, Syktus J, Deo RC, Lawrence PJ, McGowan HA, Watterson IG, Phinn SR (2007) Modeling the impact of historical land cover change on Australia's regional climate. *Geophysical Research Letters* **34**, L22711.1–L22711.6.

Michael DR, Lindenmayer DB (2008) Records of the inland carpet python, *Morelia spilota metcalfei* (Serpentes: Pythonidae), from the south-western slopes of New South Wales. *Proceedings of the Linnean Society of New South Wales* **129**, 253–261.

Michael DR, Lindenmayer DB (2018) *Rocky Outcrops in Australia: Ecology, Conservation and Management*. CSIRO Publishing, Melbourne.

Michael DR, Lunt ID, Robinson WA (2004) Enhancing fauna habitat in grazed native grasslands and woodlands: use of artificially placed log refuges by fauna. *Wildlife Research* **31**, 65–71. doi:10.1071/WR02106

Michael DR, Cunningham RB, Lindenmayer DB (2008) A forgotten habitat? Granite inselbergs conserve reptile diversity in fragmented agricultural landscapes. *Journal of Applied Ecology* **45**, 1742–1752. doi:10.1111/j.1365-2664.2008.01567.x

Michael DR, Lindenmayer DB, Cunningham RB (2010) Managing rock outcrops to improve biodiversity conservation in Australian agricultural landscapes. *Ecological Management & Restoration* **11**, 43–50. doi:10.1111/j.1442-8903.2010.00512.x

Michael DR, Cunningham RB, Lindenmayer DB (2011) Regrowth and revegetation in temperate Australia presents a conservation challenge for reptile fauna in agricultural landscapes. *Biological Conservation* **144**, 407–415. doi:10.1016/j.biocon.2010.09.019

Michael DR, Wood JT, Crane M, Montague-Drake R, Lindenmayer DB (2014) How effective are agri-environment schemes for protecting and improving herpetofaunal diversity in Australian endangered woodland ecosystems? *Journal of Applied Ecology* **51**, 494–504. doi:10.1111/1365-2664.12215

Michael DR, Ikin K, Crane M, Okada S, Lindenmayer DB (2017) Scale-dependent occupancy patterns in reptiles across topographically different landscapes. *Ecography* **40**, 415–424. doi:10.1111/ecog.02199

Montague-Drake RM, Lindenmayer DB, Cunningham RB (2009) Factors affecting site occupancy by woodland bird species of conservation concern. *Biological Conservation* **142**, 2896–2903. doi:10.1016/j.biocon.2009.07.009

Montague-Drake R, Lindenmayer DB, Cunningham RB, Stein J (2011) A reverse keystone species affects the landscape distribution of woodland avifauna: a case study using the Noisy Miner (*Manorina melanocephala*) and other Australian birds. *Landscape Ecology* **26**, 1383–1394. doi:10.1007/s10980-011-9665-4

Munro N, Lindenmayer DB (2011) *Planting for Wildlife: A Practical Guide to Restoring Native Woodlands.* CSIRO Publishing, Melbourne.

Munro N, Fischer J, Wood J, Lindenmayer DB (2009) Revegetation in agricultural areas: the development of structural complexity and floristic diversity. *Ecological Applications* **19**, 1197–1210. doi:10.1890/08-0939.1

Ng K, Driscoll DA, Macfadyen S, Barton P, McIntyre S, Lindenmayer DB (2017) Contrasting beetle assemblage responses to cultivated farmlands and native woodlands in a dynamic agricultural landscape. *Ecosphere* **8**(1), e02042. doi:10.1002/ecs2.2042

Ng K, McIntyre S, Barton PS, Macfadyen S, Driscoll DA, Lindenmayer DB (2018) Dynamic effects of ground-layer plant communities on beetles in a fragmented farming landscape. *Biodiversity and Conservation.* doi:10.1007/s10531-018-1526-x

Prober SM, Thiele KR, Lunt ID, Koen TB (2005) Restoring ecological function in temperate grassy woodlands: manipulating soil nutrients, exotic annuals and native perennial grasses through carbon supplements and spring burns. *Journal of Applied Ecology* **42**, 1073–1085. doi:10.1111/j.1365-2664.2005.01095.x

Pulsford SA, Driscoll DA, Barton PS, Lindenmayer DB (2017a) Remnant vegetation, plantings and fences are beneficial for reptiles in agricultural landscapes. *Journal of Applied Ecology* **54**, 1710–1719. doi:10.1111/1365-2664.12923

Pulsford SA, Lindenmayer DB, Driscoll DA (2017b) Reptiles and frogs conform to multiple conceptual landscape models in an agricultural landscape. *Diversity and Distributions* **23**, 1408–1422. doi:10.1111/ddi.12628

Reid N (1991) Coevolution of mistletoes and frugivorous birds? *Australian Journal of Ecology* **16**, 457–469. doi:10.1111/j.1442-9993.1991.tb01075.x

Robinson D (2006) Is revegetation in the Sheep Pen Creek Area, Victoria, improving Grey-crowned Babbler habitat? *Ecological Management & Restoration* **7**, 93–104. doi:10.1111/j.1442-8903.2006.00263.x

Romanowski R (2009) *Planting Wetlands and Dams.* 2nd edn. Landlinks Press, Melbourne.

Sato C, Wood J, Stein JA, Crane M, Okada S, Michael D, Kay G, Florance D, Seddon J, Gibbons P, Lindenmayer DB (2016) Natural tree regeneration in agricultural landscapes: the implications of intensification. *Agriculture, Ecosystems & Environment* **230**, 98–104. doi:10.1016/j.agee.2016.05.036

Selwood K, Mac Nally R, Thomson JR (2009) Native bird breeding in a chronosequence of revegetated sites. *Oecologia* **159**, 435. doi:10.1007/s00442-008-1221-9

Simpson P (1993) *Productivity of Native Grasses-based Pasture on the Tablelands.* NSW Department of Agriculture, Sydney.

Smallbone LT, Prober SM, Lunt ID (2007) Restoration treatments enhance early establishment of native forbs in degraded temperate grassy woodland. *Australian Journal of Botany* **55**, 818–830. doi:10.1071/BT07106

Spooner PG, Briggs SV (2008) Woodlands on farms in southern

New South Wales: a longer-term assessment of vegetation changes after fencing. *Ecological Management & Restoration* **9**, 33–41. doi:10.1111/j.1442-8903.2008.00385.x

Spooner P, Lunt I, Robinson W (2002) Is fencing enough? The short-term effects of stock exclusion in remnant grassy woodlands in southern NSW. *Ecological Management & Restoration* **3**, 117–126. doi:10.1046/j.1442-8903.2002.00103.x

Steffen W, Burbidge A, Hughes L, Kitching R, Lindenmayer DB, Musgrave W, Stafford-Smith M, Werner P (2009) *Australia's Biodiversity and Climate Change.* CSIRO Publishing, Melbourne.

Stirzaker R, Vertessey R, Sarre A (Eds) (2002) *Trees, Water and Salt: An Australian Guide to Using Trees for Healthy Catchments and Productive Farms.* Joint Venture Agroforestry Program, Canberra.

Streatfield S, Fifield G, Pickup M (2010) Whole of paddock rehabilitation (WOPR): a practical approach to restoring Grassy Box woodlands. In *Temperate Woodland Conservation and Management.* (Eds DB Lindenmayer, AF Bennett, RJ Hobbs) pp. 23–31. CSIRO Publishing, Melbourne.

Thomas R (2009) Regent Honeyeater habitat restoration project Lurg hills, Victoria. *Ecological Management & Restoration* **10**, 84–97. doi:10.1111/j.1442-8903.2009.00470.x

Thomson LJ, Hoffmann AA (2010a) Cost benefit analysis of shelterbelt establishment: natural enemies can add real value to shelterbelts. *The Australian & New Zealand*

Grapegrower & Winemaker **554**, 38–44.

Thomson LJ, Hoffmann AA (2010b) Natural enemy responses and pest control: importance of local vegetation. *Biological Control* **52**, 160–166. doi:10.1016/j.biocontrol.2009.10.008

Thomson LJ, Hoffmann AA (2013) Spatial scale of benefits from adjacent woody vegetation on natural enemies within vineyards. *Biological Control* **64**, 57–65. doi:10.1016/j.biocontrol.2012.09.019

Tulloch AIT, Mortelliti A, Kay G, Florance D, Lindenmayer DB (2016) Using empirical models of species colonization under multiple threatening processes to identify complementary threat-mitigation strategies. *Conservation Biology* **30**, 867–882. doi:10.1111/cobi.12672

van der Ree R, Bennett AF, Gilmore DC (2004) Gap-crossing by gliding marsupials: thresholds for use of isolated woodland patches in an agricultural landscape. *Biological Conservation* **115**, 241–249. doi:10.1016/S0006-3207(03)00142-3

Vesk P, Nolan R, Thomson JW, Dorrough JW, Mac Nally R (2008) Time lags in the provision of habitat resources through revegetation. *Biological Conservation* **141**, 174–186. doi:10.1016/j.biocon.2007.09.010

Waldron A, Miller DC, Redding D, Mooers A, Kuhn T, Nibbelink N, Roberts JT, Tobias JA, Gittleman JL (2017) Reductions in global biodiversity loss predicted from conservation spending. *Nature* **551**, 364–367. doi:10.1038/nature24295

Walker J, Bullen F, Williams BG (1993) Ecohydrological changes in the Murray-Darling Basin. I. The number of trees cleared over two centuries. *Journal of Applied Ecology* **30**, 265–273. doi:10.2307/2404628

Walpole SC (1999) Assessment of the economic and ecological impacts of remnant vegetation on pasture productivity. *Pacific Conservation Biology* **5**, 28–35. doi:10.1071/PC990028

Walpole S, Lockwood M, Miles CA (1998) Influence of remnant native vegetation on property sale price. Johnstone Centre, Charles Sturt University, Albury.

Walters JR, Ford HA, Cooper CB (1999) The ecological basis of sensitivity of brown treecreepers to habitat fragmentation: a preliminary assessment. *Biological Conservation* **90**, 13–20. doi:10.1016/S0006-3207(99)00016-6

Watson DM (2001) Mistletoe – a keystone resource in forests and woodlands worldwide. *Annual Review of Ecology and Systematics* **32**, 219–249. doi:10.1146/annurev.ecolsys.32.081501.114024

Whittingham MJ (2007) Will agri-environment schemes deliver substantial biodiversity gain, and if not why not? *Journal of Applied Ecology* **44**, 1–5. doi:10.1111/j.1365-2664.2006.01263.x

Williams J, Price RJ (2010) Impacts of red meat production on biodiversity in Australia: a review and comparison with alternative protein production industries. *Animal Production Science* **50**, 723–747. doi:10.1071/AN09132

Woinarski JC, Burbidge AA, Harrison PL (2015) Ongoing unraveling of a continental fauna: decline and extinction of Australian mammals since European settlement. *Proceedings of the National Academy of Sciences of the United States of America* **112**, 4531–4540. doi:10.1073/pnas.1417301112

Zaradic PA, Pergams OR (2007) Videophilia: implications for childhood development and conservation. *The Journal of Developmental Processes* **2**, 130–144.

Appendix – List of scientific and common names of species named in the text

Exotic species are denoted with a star (*).

	Common name	Scientific name
Mammals	Bush Rat	*Rattus fuscipes*
	Common Brushtail Possum	*Trichosurus vulpecula*
	Common Ringtail Possum	*Pseudocheirus peregrinus*
	Dingo	*Canis lupus dingo*
	Eastern Grey Kangaroo	*Macropus giganteus*
	European Rabbit	*Oryctolagus cuniculus*
	Feral Cat	*Felis catus*
	Greater Glider	*Petauroides volans*
	Grey-headed Flying Fox	*Pteropus poliocephalus*
	Little Red Flying Fox	*Pteropus scapulatus*
	Marsupial Lion	*Thylacoleo* spp.
	Red Fox	*Vulpes vulpes*
	Squirrel Glider	*Petaurus norfolcensis*
	Sugar Glider	*Petaurus breviceps*
	Tasmanian Tiger	*Thylacinus cynocephalus*
	Yellow-footed Antechinus	*Antechinus flavipes*
Birds	Australian Ibis	*Threskiornis moluccus*
	Australian Magpie	*Cracticus tibicen*
	Brown Treecreeper	*Climacteris picumnus*
	Bush Stone-curlew	*Burhinus grallarius*
	Clamorous Reed Warbler	*Acrocephalus stentoreus*
	Common Starling*	*Sturnus vulgaris*
	Dusky Woodswallow	*Artamus cyanopterus*
	European Goldfinch	*Carduelis carduelis*
	Flame Robin	*Petroica phoenicea*
	Glossy Black-cockatoo	*Calyptorhynchus lathami*
	Grey Butcherbird	*Cracticus torquatus*
	Grey-crowned Babbler	*Pomatostomus temporalis*
	Grey Shrike-thrush	*Colluricincla harmonica*
	House Sparrow*	*Passer domesticus*
	Noisy Miner	*Manorina melanocephala*
	Rufous Songlark	*Megalurus mathewsi*
	Rufous Whistler	*Pachycephala rufiventris*
	Southern Whiteface	*Aphelocephala leucopsis*
	Speckled Warbler	*Pyrrholaemus sagittatus*
	Superb Fairy-wren	*Malarus cyaneus*
	Swift Parrot	*Lathamus discolor*
	Wedge-tailed Eagle	*Aquila audax*
	White-browed Babbler	*Pomatostomus superciliosus*
	White-plumed Honeyeater	*Lichenostomus penicillatus*
	White-winged Triller	*Lalage tricolor*

(Continued)

Appendix (Continued)

	Common name	Scientific name
Reptiles	Bearded Dragon	*Pogona* spp.
	Boulenger's Skink	*Morethia boulengeri*
	Carpet Python	*Morelia spilota*
	Curl Snake	*Suta suta*
	Eastern Brown Snake	*Pseudonaja textilis*
	Olive Legless Lizard	*Delma inornata*
	Southern Rainbow Skink	*Carlia tetradactyla*
	Striped Legless Lizard	*Delma impar*
	Three-toed Earless Skink	*Hemiergis decresiensis*
	Tessellated Gecko	*Diplodactylus tessellatus*
	Wall Skink	*Cryptoblepharus virgatus*
Plants	Bulbine Lily	*Bulbine bulbosa*
	Kangaroo Grass	*Themeda australis*
	Manna Gum	*Eucalyptus viminalis*
	Mugga Ironbark	*Eucalyptus sideroxylon*
	She-oak	*Casuarina* spp.
	Variable Plantain	*Plantago varia*
	Wattle	*Acacia* spp.
	White Box	*Eucalyptus albens*
	Ribbon Gum	*Eucalyptus viminalis*
	Yam Daisy	*Microseris lanceolata*
	Yellow Gum	*Eucalyptus leucoxylon*

Index

Note: Bold page numbers refer to illustrations.